FIFTY YEARS OF
GENETIC LOAD

BY THE SAME AUTHOR

Chromosomes, Giant Molecules, and Evolution
Topics in Population Genetics
Essays in Social Biology, 3 vols.
Basic Population Genetics

COAUTHOR

Radiation, Genes, and Man
Adaption
Biology for Living

EDITOR

Human Culture: A Moment in Evolution
Dobzhansky's Genetics of Natural Populations I–XLIII (4th editor)

FIFTY YEARS OF GENETIC LOAD

An Odyssey

Bruce Wallace

Cornell University Press

ITHACA AND LONDON

Library of Congress Cataloging-in-Publication Data
Wallace, Bruce.
  Fifty years of genetic load : an odyssey / Bruce Wallace.
    p. cm.
  Includes bibliographical references and index.
  ISBN 0-8014-2583-2 (alk. paper)
  1. Genetic load. 2. Population genetics. I. Title. II. Title:
50 years of genetic load.
QH455.W35 1991
575.1'5—dc20   90-55754

Copyright © 1991 by Cornell University

All rights reserved. Except for brief quotations in a review, this book, or parts thereof, must not be reproduced in any form without permission in writing from the publisher. For information, address Cornell University Press, 124 Roberts Place, Ithaca, New York, 14850.

First published 1991 by Cornell University Press.

Printed in the United States of America

∞ The paper in this book meets the minimum requirements of the American National Standard for Information Sciences—Permanence of Paper for Printed Library Materials, ANSI Z39.48-1984.

In memory of Terumi Mukai

*Terumi was a friend whose
work complemented mine
in many ways, for many years.*

CONTENTS

	Preface	ix
1	Introduction	1
2	Genetic Variation, Darwinian Fitness, and Genetic Load	8
3	Studies of Irradiated Populations	23
4	Random Mutations and Viability	49
5	Dilemmas and Options	68
6	Hard and Soft Selection	80
7	Persistence: An Important Component of Population Fitness	104
8	Self-culling and the Persistence of Populations	111
9	Summarizing Remarks	133
10	Still to Come . . .	139
	References	161
	Index	171

PREFACE

When Theodosius Dobzhansky's *Genetics of the Evolutionary Process* (the final edition of several that were initially titled *Genetics and the Origin of Species*) appeared in 1970, one reviewer confessed to a sigh of relief. At the outset, he had been concerned that Dobzhansky, in attempting to make a final revision at an advanced age, would produce a parody of the text that had been the cornerstone of the Modern Synthesis. His worries, the reviewer learned, had been baseless. Perhaps they can now be reawakened by my attempt to review fifty years of genetic load; I can only hope that the same reviewer's fears will prove baseless once more.

Although I am a population geneticist, my approach to the study of populations has not been traditional. I have never run a computer simulation. I have not studied allozyme variation extensively, nor, outside the classroom, have I ever calculated a genetic distance. I have never indulged in mathematics much higher than high school algebra.

The nontraditional nature of my research is explained, I think, by an eleven-year sojourn at Cold Spring Harbor, New York, that began as I left Dobzhansky's laboratory at Columbia University. At weekly staff meetings I had to report on my research to an audience that included Milislav Demerec (director of both the Biological Laboratory at Cold Spring Harbor and the Department of Genetics of the Carnegie Institution of Washington), A. D. Hershey, Barbara McClintock, and Evelyn

Preface

Witkin. None of these persons was a population geneticist, but each had an intense interest in science, and each possessed a tremendous intellect. My task was to state my current research problem clearly and to describe the procedures by means of which I intended to solve it. If I succeeded in my report, every comment from the audience—no matter how casually made—was a valuable one requiring at times hours of thought.

One issue regarding the use of Hardy-Weinberg equilibria (an issue that was subsequently addressed by four authors in correspondence to *Evolution* at the time and that has been resurrected by still another within the 1980s) was summarized by Hershey as follows: "If you know where you are but you do not know where you started, then you cannot say how you got there." True! One can only guess at the most likely path.

Summers and the annual Cold Spring Harbor symposia brought still larger audiences to be addressed; among the regular visitors were Max Delbrück, S. E. Luria, Guido Pontecorvo, Leo Szilard, Ernst Caspari, and (then a young graduate student) J. D. Watson. For these persons—most of whom (with the notable exception of Barbara McClintock) were phage and microbial geneticists—research problems had to be precise if they were to be worthwhile, and laboratory procedures aimed at solving them had to involve the fewest possible assumptions. I recall with some pride a local staff member telling me one June afternoon that mine was the only symposium paper of the day that dealt with a clearly defined problem and involved reasoning and research procedures that he could understand.

A second memory is of a morning in conversation with Delbrück, at his insistence; he was preparing himself for the task of teaching general biology at Cal Tech—a task for which he had volunteered. Our talk got around to the fate of a single mutant gene in a large diploid population. Knowing that each pair of parents must be replaced by two offspring, I said that the probability of losing the gene from the population was $\frac{1}{2} \times \frac{1}{2}$, or $\frac{1}{4}$. Delbrück was surprised at this; he cited the Poisson distribution and claimed that the probability of loss would be 0.37. He would not let me concede error, however, and move on to another topic. We were both correct, he explained, and therefore there must be an explanation for our different answers. Shortly, the explanation occurred to him: my calculation had removed variation in numbers of progeny per pair of parents. Subsequently, I noticed that both solutions to this problem had been included among the chapter-end exercises of one widely used genetics textbook.

The environment at Cold Spring Harbor was not one that encouraged an extensive use of quantitative genetics; had I gone (as I might have) to North Carolina State University in the mid-1950s, I would have joined S. G. Stephens, H. F. Robinson, R. E. Comstock, and C. C. Cockerham, and my subsequent research would undoubtedly have been profoundly influenced by these good friends.

There were no ecologists at Cold Spring Harbor; hence, I had no opportunity to take part in the creation of a population biology as did, for example, R. C. Lewontin, Robert MacArthur, Richard Levins, and E. O. Wilson (four Young Turks in 1968, when Lewontin edited the symposium volume *Population Biology and Evolution*) through their collaboration with one another and with other university colleagues. My contact with ecologists was postponed until my arrival at Cornell University, where Simon Levin, Richard Root, and William Brown became welcome colleagues. My participation in the symposium that Lewontin arranged at Syracuse University in 1967 was entirely accidental; I had sent him a manuscript requesting criticism and comments. His response was an invitation to present the material at Syracuse, hence my participation in, or at least my presence at, the creation of population biology.

Because all research at the Biological Laboratory during the 1950s was supported by "soft" money, the pace was not the leisurely one generally associated with university departments. When asked how the staff at the Biological Laboratory could accomplish so much, Vernon Bryson, a colleague, responded that when next year's salary depends upon this year's accomplishments, one works harder. This pace discouraged the use of students to perform what was essentially contract research. During special experiments called "big pushes," half the laboratory technicians would work the usual 8 A.M.–5 P.M. shift, and the remaining half would work from noon until the day's work ended near 10 P.M. During such times my colleague J. C. King and I would work from 8 A.M. until near midnight. A two-day holiday would be declared at the termination of such an experiment; New York State's Fair Employment Practices Act had not yet become law.

The account given here of the influence those eleven years at Cold

Preface

Spring Harbor had on my research efforts may explain my unconventionality, but it does not excuse it, if, in fact, it needs excusing. In retrospect, I feel that each phase of my research has more or less set the stage for the next one. There has been a continuity that, in my view, has conferred an overall pattern on my work; this pattern includes a gradual change in my point of view—a change based on what I consider to be a series of rational judgments. I can imagine, in contrast, having spent my life studying empirical relationships between genetic markers and some physiological trait in natural populations in which existing complexities would have precluded predictions. After a lapse of several decades, a younger worker might repeat my study and find that the relationships I had observed no longer held. At that time, I would probably publish a joint paper with the younger person, dealing with and speculating on the reversal of a longtime correlation. To what purpose? This is precisely the type of study that would have palled at Cold Spring Harbor—one that my then colleagues would have either ignored or vigorously criticized. I greatly appreciate the guiding influence that those marginally interested parties had on my early scientific development, largely by example but also by their insightful comments on my research.

In concluding these remarks, I wish to thank the Alexander von Humboldt–Stiftung for the Senior U.S. Scientist Award (1986 and 1987), which made writing this book possible. Professor Diether Sperlich, my host at the University of Tübingen, and his colleagues provided a most gracious and stimulating environment within which to work; I thank them for their many kindnesses. Several colleagues read and commented encouragingly on the penultimate draft of this manuscript: John Beatty of the University of Minnesota, Robert May of Oxford University, Ernst Mayr of Harvard University, and Will Provine of Cornell University. I owe them all many thanks, including the one who observed that this "is by no means a balanced treatment by a disinterested observer." I concede that he is quite right: this is an account of *my* odyssey in which I emphasize the floes, whirlpools, reefs, and shoals that threatened me during *my* voyage.

BRUCE WALLACE

Blacksburg, Virginia

FIFTY YEARS OF
GENETIC LOAD

1

INTRODUCTION

The synthetic theory of evolution (also known as the Modern Synthesis) began sometime before 1937, the year I entered Columbia College. At least, Dobzhansky's *Genetics and the Origin of Species* appeared in that year. The synthetic theory represented a major revolution in evolutionary thought; Ernst Mayr (1982) claims that it was essentially completed by 1947, thus ending a decade that included the publication (or preparation) of three additional classics: Mayr's (1942) *Systematics and the Origin of Species*, G. L. Stebbins's (1950) *Variation and Evolution in Plants*, and G. G. Simpson's (1944) *Tempo and Mode in Evolution*. These four authors represented genetics, animal and plant systematics, and paleontology. Each, of course, also represented a host of earlier workers whose contributions were melded into the final synthesis. This is especially true in the case of genetics. The foundations of population genetics had been laid during the 1920s and early 1930s by R. A. Fisher (1930), J. B. S. Haldane (1932), and Sewall Wright (1931). Dobzhansky's contribution to the Modern Synthesis was, consequently, itself a synthesis of theoretical population genetics and his own superb knowledge of the genetics and ecology of natural populations.

The synthetic theory, based as it is on many biological disciplines, encompasses identifiable subtheories. One of these is the theory of genetic loads. Although the term *genetic load* was first used by H. J.

Fifty Years of Genetic Load

Miller in 1950, the essential concept appears in Haldane's (1937) paper titled "The Effect of Variation on Fitness." In referring to fifty years of genetic load, I use Haldane's, rather than Muller's, paper as a starting point.

Three decades later, the neutral theory was outlined by Motoo Kimura (1968). Although related to Wright's (1931) notion of genetic drift, the neutral theory has matured into an alternative to the synthetic theory (see Kimura, 1983). In the chapters that follow, the discussion will of necessity touch on the neutral theory, but only so far as that theory impinges directly on the concept of genetic loads.

Studies of evolution involve different facets of living and extinct organisms. These facets or levels might be compared with the facets or levels that one would encounter in studying the evolution of a university. Architects, engineers, and urban planners, for example, might be interested in plotting the temporal sequences and spatial patterns with which buildings have appeared and disappeared on a university's campus, and with the growth of the campus itself. Presumably, their account of a university would begin with a description of the original "Old Main"—a building that housed the first administrators and the earliest classrooms—and would continue to the present-day complex of theaters, classroom buildings, science laboratories, and sports complexes.

Educators and those interested in the history of education might ignore the university's physical plant while compiling data on the evolution of its educational programs: perhaps electrical engineering was first offered in 1875, advanced degrees in genetics in 1938, and theater arts in 1969. With some justification, these persons could claim that they are studying the most important aspect of the university, at least to the extent that education is the main function of any university. Educational programs, however, are not independent of physical facilities; classes in electrical engineering cannot be profitably held in a botanical garden, nor can theater rehearsals be held in small classrooms. Consequently, those studying the evolution of the physical plant and those compiling information on past and present educational programs will note many correlations between their observations. Generally, they will note that programs have demanded facilities; on occasion, however, they should find that the presence of a

Introduction

building permitted the development of a particular instructional program—generally entailing minor, but necessary, physical alterations.

The evolution of a university can be studied at still a third level: past personnel records could be used to chart the composition of the staff for each year following the university's inception. This level is analogous to the molecular level of evolutionary studies. For the most part, the faculty would appear to change in a random fashion. A specialist in European history retires, but within the year he is replaced by another person; the curriculum continues seemingly unchanged. Faculty members depart to accept positions elsewhere, retire, or die; as a rule, they are quickly replaced. Occasionally, a professor develops a special program that is reflected in an altered curriculum and eventually, perhaps, even in the university's physical plant. Such events are rare, however. Indeed, one might easily devise a neutral theory of faculty composition.

If the main purpose of a university is the education of its students, then its faculty—the physical bodies who instruct, interpret information, and judge when the student's education is complete—will be regarded by many as the most important facet or level in the task of educating. Nevertheless, a historical tabulation of the comings and goings, the granting and nongranting of tenure, and the seemingly instantaneous replacement of retirees appears to be a tabulation of transient, virtually random events. Analogies should not be carried too far, of course, but a further comment can be ventured: the appearance of randomness in faculty composition exists despite the elaborate systems of interviews and seminars that most universities rely upon in selecting new faculty members. That *selection* has operated at any time in determining the composition of a university's faculty is not emphasized in the historical roster of its staff members. Paraphrasing J. L. King and T. H. Jukes (1969), one might say that evolutionary change at the physical and curricular levels of the university results from the process of selection operating through an adaptive change in faculty composition. It does not necessarily follow that all, or even most, changes of faculty are the result of such selection. (The latter sentence would incur the displeasure of most faculty search committees!)

In addition to genetic load theory and the neutral theory, the synthetic theory has encompassed a number of named suggestions or ideas; among these would be "genetic assimilation" (Waddington 1953, 1956), "coadaptation" (Dobzhansky, 1950; Wallace and Vetukhiv, 1955), "the founder theory" (Mayr, 1954), "genetic homeostasis" (Lerner, 1954), and others. Kimura (1983:22) listed these lesser ideas as examples illustrating the overdevelopment of the synthetic theory—its Baroque period, so to speak. Interestingly, E. O. Wilson (1975) has also compiled a list of terms to which he objects: fitness, genetic drift, gene migration, and mutation pressure are cited specifically. Both Kimura and Wilson are objecting, I believe, to the use of words as, or the substitution of words for, explanations. Words do not explain. (A patient with a rapid pulse learns nothing from the physician's diagnosis of tachycardia.) Furthermore, phenomena (that is, the terms used to identify these phenomena) that are still to be proved are often advanced as explanations. Even now, nearly a century after the rediscovery of Mendel's paper, inconclusive numerical data (those, for example, that approximate 3 : 1, 15 : 1, 63 : 1, or still other ratios that can be generated by Mendelian systems) are frequently advanced as *genetic* evidence. Now, if Mendelian inheritance must still be proved in individual instances after so many years, then unproven suggestions (complex suggestions, at that) of more recent origin mut be used as explanations only with extreme care.

An important point, however, may be overlooked by those who criticize the use of neologisms and other terms of explanations: each such term when it was first proposed referred to observations that required interpretation. Each term originally was merely a label for a proffered interpretation of a particular set of observations. Criticism of the subsequent abuse of a term does not remove the necessity for devising an explanation for the original observations. Inconveniently discordant data, that is, cannot be dismissed by merely criticizing terms; on the contrary, the data themselves demand explanation.

The account of the genetic load theory given in the following chapters is largely a personal account of my encounter with, studies of, and reactions to this theory. If the geometry weren't wrong, the title of this book would have been *Closing the Ring: Fifty Years of Genetic Load*. Unfortunately, the account proceeds from my acceptance that a load (genetic or phenotypic) harms a population, to a later belief that it has little or no bearing on a population's well-being, to my present feeling that a phenotypic load (which may or may not have a genetic basis) provides the means for the culling of excess individuals, thus avoiding overcrowding and increasing the probability that a population will persist through time. Rather than having closed a ring, I now appear to be 180° from my starting point. The steps by which I have arrived at my present views will be retraced in the chapters of this book. These steps, I believe, fall on the path that leads to the eventual (but, today, still imperfect) union of population genetics and ecology into a single science: *population biology*.

*Personal comments**

A reviewer of this manuscript has urged me to "make clearer the central importance of the genetic load controversies to the history of recent evolutionary biology." That there were such controversies has been confirmed by a recent conversation with an elderly colleague who asked whether I remembered the great battles of the fifties and sixties. Such joggings conjure up a variety of remembrances.

One is of a Brazilian geneticist who, with tongue in cheek, remarked during the 1960s that a great scientist is one who can sustain a great controversy. Like many trivial events, that remark has had a lasting effect: the Dobzhansky-Muller exchanges concerning genetic load have become inextricably entangled with my memories of the Jack Benny–Fred Allen feud of early radio. Much of the fame of these two men rested on their weekly exchanges of sarcastic jibes.

A second incident that I clearly recall is my attempt one night (perhaps in 1963) to explain the Dobzhansky-Muller controversy to my wife—at her request, since she had also been a Dunn-Dobzhansky student at Columbia University. The more I explained, the more ridiculous the whole matter seemed in comparison with problems the family (a daughter's encounter with pencillin-resistant staphylococcus), the country (events culminating in President Kennedy's assassination were either brewing or had already exploded), and the world (many persons

---

* At the end of each chapter I shall make comments of a personal nature, frequently in the form of anecdotes which are relevant to but which have only an ancillary bearing on the subject under discussion. The neutralist-selectionist controversy mentioned here is with us still: should "survival of the luckiest" replace "survival of the fittest" as the underpinning for organic evolution?

were lamenting the worsening quagmire that was Southeast Asia) were experiencing. As I dealt with each contentious issue in increasingly mock seriousness, the two of us were reduced to uncontrolled, hysterical laughter. The genetic load controversy that evening assumed, in my judgment, its proper dimensions.

I do not wish to be irreverent, but the attraction and joy of science as an occupation, in my opinion, is recognizing questions that beg for solutions, devising means for arriving at those solutions, and doing what must be done to obtain the answers. Seeing how to solve a problem is more exciting, of course, than performing the actual test; the latter task may involve sheer drudgery.

When two scientists of goodwill differ, they never do so entirely. Their views probably coincide almost completely. After testing out one another's views, they may discover that in 99 consecutive instances both have responded yes or no in unison. They then find at the 100th instance that one says yes and the other, no. The two may even run through the preliminaries again, and still again, only to find that after seemingly total agreement, they cannot concur at the conclusion. The resulting frustration is a large part of the controversy that then develops.

Among other possibilities, the point of disagreement—the 100th item—may lie beyond the current realm of experimentation. Or, it may reflect a difference in either viewpoint or vocabulary. H. J. Muller, for example, by seemingly impeccable logic arrived at the conclusion that for a given species there is one, and only one, normal allele at each gene locus; all alleles that differ from this one must to some degree be detrimental to the health and well-being of their carriers. J. C. King and I, on the other hand, noted that advances in experimental techniques consistently revealed variation where uniformity had previously prevailed: genetic tests had revealed large stores of heritable variation within populations of seemingly uniform *Drosophila* flies; giant salivary gland chromosomes in the Diptera had revealed a wealth of chromosomal inversions that had not been detected by examining mitotic chromosomes in other tissues; and serological advances were subdividing simple blood group systems (the A-B-O system, for example) into more complex ones. We saw no reason to believe that this trend had run its course. Here was a difference of opinion. Having expressed it, however, little was to be gained by further belaboring the

Introduction

matter; it would eventually be settled by technical advances. And the matter has been resolved: variation is rampant at the gene level. Population geneticists now safely assume that each new mutation results in an allele unlike any other currently existing in the population.

Neutrality (of which we shall hear more in later chapters) is a matter that may depend as much upon one's perspective as upon "facts." Survival and successful reproduction are complex events that represent the sum of many individual components: avoiding predators, obtaining food, finding a mate, and producing and (perhaps) rearing offspring. Each of these components in turn represents the outcome of a lengthy series of events. There are, for example, many factors involved in not being eaten: behavior, speed, size, defense, camouflage, and more. Referring to survival and successful reproduction as *fitness,* one sees that there must be little or no correlation between the individual elements upon which fitness ultimately depends and fitness itself. The correlation between the sum of many (n) equally important elements and the individual elements equals $\sqrt{1/n}$ if the elements are not correlated with one another (Reeve et al., 1990). If the elements are correlated (γ) (speed with size, for example), the correlation between the sum and the individual elements equals $\sqrt{[(1 - \gamma)/n] + \gamma}$. The latter relationship reveals that the correlation between fitness and its individual components equals zero if the individual components exhibit even a small $[-1/(n - 1)]$ negative correlation. Now, it is known that within equilibrium populations various components of fitness are indeed negatively correlated. For example, rapid development generally leads to small size and, in the case of females, to a corresponding reduction in egg number or size, or both. Avoidance of predators is sometimes accomplished only by forgoing food and water, items essential for survival. At this point, I leave it to the reader to decide whether a phenotypic trait that is an obvious component of fitness but exhibits no correlation with fitness itself is a neutral trait and, if so, under what definition of neutrality. Here, I suspect, are to be found the roots of a controversy. In my opinion, the facts—the realities of the situation— are much more exciting than the particular words that different persons use in describing their different points of view. Each of us is blind to some degree, and life is an extremely complex elephant. My joy has come from exploring, rather than describing, that part of the elephant which I have chanced to encounter.

2

GENETIC VARIATION, DARWINIAN FITNESS, AND GENETIC LOAD

Haldane's (1937) paper titled "The Effect of Variation on Fitness" marks the origin of the genetic load concept. In this paper Haldane emphasized the relationship between mutation and the average fitness of a population; earlier workers (e.g., S. Wright, 1931) had stressed instead the equilibrium gene frequencies that result from the opposing actions of mutation and selection.

The results obtained by Haldane will be illustrated by calculations patterned after those used by J. F. Crow (1948). First, though, zygotic frequencies and Darwinian fitnesses will be combined in calculating the average Darwinian fitness of a population; this average itself requires at least a brief discussion.

Throughout this chapter, illustrations will be based on a single locus that is occupied by two alleles, A and a, whose frequencies are p and q, where $p + q = 1.00$. The expected (i.e., Hardy-Weinberg) proportions of AA, Aa, and aa individuals are p^2, $2pq$, and q^2, respectively.

The Darwinian fitness of an individual is a measure of the reproductive success of that individual relative to the average reproductive success of the other members of the population. As a rule, Darwinian fitness refers to differences in reproductive success that have a heritable component and therefore result in genetic changes within all but equilibrium populations. Such changes are regarded as the most elementary of all evolutionary changes: *microevolution*.

Variation, Fitness, and Genetic Load

The Darwinian fitness of a single individual has little bearing on evolutionary change; too many accidents are responsible for the observed value. A young, potentially vigorous and fecund insect, for example, may be destroyed by the actions of a less-than-perfect but older one. Because of such accidentally determined variation in individual fitnesses, Darwinian fitnesses (actually, *average* Darwinian fitnesses) are most often assigned to individuals representing various *genotypes*. The standard format for this procedure can be illustrated as follows:

Genotype	AA	Aa	aa
Frequency	p^2	$2pq$	q^2
Fitness	1	1	$1 - s$
Frequency after selection	p^2	$2pq$	$q^2 - sq^2$
Average fitness ($\overline{W}$)		$1 - sq^2$	

In this tabulation, the value 1.00 is assigned to the genotype (i.e., to the average fitness of individuals of that genotype) exhibiting the greatest success in leaving fertile adult offspring. If egg production determines the number of progeny produced, the fitnesses 1, 1, and $1 - s$ could refer to 139, 139, and 107 eggs as the average numbers produced by AA, Aa, and aa females; if so, $s = 0.23$. Fitness consists of numerous components of which egg production (fecundity) is only one; others include survival, mating activity, longevity, and fertility. As a practical matter, fitness is extremely difficult, perhaps impossible, to measure directly; only components of fitness can be identified and then measured experimentally. The interpretation of such empirical data is difficult: they are often gathered under experimental conditions differing from those prevailing within the population under study; in addition, appropriate weighting factors (for longevity, for example) are generally unknown.

The mathematical treatment of biological problems is accomplished only with the aid of explicit and implicit assumptions. The tabulation presented above is no exception. Because selection is said to act in this case after zygote formation (fertilization) and before sexual maturity (reproduction), the selective disadvantage of aa individuals is assigned to these individuals alone. The confidence with which this assignment is made can be misleading: the inviability (or other shortcoming) of

Fifty Years of Genetic Load

	AA p^2 1	Aa $2pq$ 1	aa q^2 1-s
AA p^2 1	p^4 AA	p^3q AA p^3q Aa	$(1-s)p^2q^2$ Aa
Aa $2pq$ 1	p^3q AA p^3q Aa	p^2q^2 AA $2p^2q^2$ Aa p^2q^2 aa	$(1-s)pq^3$ Aa $(1-s)pq^3$ aa
aa q^2 1-s	$(1-s)p^2q^2$ Aa	$(1-s)pq^3$ Aa $(1-s)pq^3$ aa	$(1-s)^2q^4$ aa

Figure 2-1. The distribution of fitnesses among individuals of different genotypes within a population following the mating of males and females whose frequencies (and relative fitnesses) are given as *AA:* p^2 (1); *Aa:* $2pq$ (1); and *aa:* q^2 $(1-s)$. Because the interactions of parental fitnesses are shown as multiplicative, the average fitness of the progeny population has decreased: $(1-sq^2)^2$. That decrease is unimportant. Once they have been assigned responsibility for generating ill-adapted homozygous progeny (as the diagram does), the ratio of the contributions to the reduced average fitness of the population made by heterozygotes and homozygotes equals p/q. If q is small, the primary responsibility for producing ill-adapted offspring can be assigned to seemingly "normal" heterozygotes.

progeny individuals can be viewed as lowering parental fertility. Who, one may ask, is responsible for the production of the selectively impaired *aa* homozygotes?

Figures 2-1 and 2-2 show that heterozygotes must share responsibility in lowering the average fitness of a population; the onus need not be restricted to the selectively inferior *aa* individuals alone. Because relative fitnesses are said to interact in a multiplicative manner (Figure 2-1), the average fitness of the population has been lowered from $1-sq^2$ to $(1-sq^2)^2$. That aspect of the figure is not important because the average fitness of a population is merely the average of relative fitnesses. Of greater importance is the relative size of the contributions of heterozygotes (*Aa*) and homozygotes (*aa*) to the lowering of the average. In both the multiplicative and additive (Figure 2-2) cases, the ratio of the contributions of heterozygotes and homo-

Variation, Fitness, and Genetic Load

	AA p^2 1	Aa $2pq$ 1	aa q^2 $1-s$
AA p^2 1	p^4 AA	p^3q AA p^3q Aa	$(1-\frac{s}{2}) p^2q^2$ Aa
Aa $2pq$ 1	p^3q AA p^3q Aa	p^2q^2 AA $2p^2q^2$ Aa p^2q^2 aa	$(1-\frac{s}{2}) pq^3$ Aa $(1-\frac{s}{2}) pq^3$ aa
aa q^2 $1-s$	$(1-\frac{s}{2}) p^2q^2$ Aa	$(1-\frac{s}{2}) pq^3$ Aa $(1-\frac{s}{2}) pq^3$ aa	$(1-s)q^4$ aa

Figure 2-2. The distribution of fitnesses among individuals of different genotypes within a population following the matings of males and females whose frequencies (and relative fitnesses) are given as *AA:* p^2 (1); *Aa:* $2pq$ (1); and *aa:* q^2 $(1-s)$. Because the interactions are shown as additive, the average fitness of the progeny population (unlike the case illustrated in Figure 2-1) remains $1 - sq^2$. As before, that is unimportant. Once they have been assigned responsibility for generating ill-adapted homozygous progeny, the ratio of the contributions to the reduced average fitness of the population made by heterozygotes and homozygotes equals p/q (as in Figure 2-1). If q is small, the primary responsibility for producing ill-adapted offspring can be assigned to seemingly "normal" heterozygotes.

zygotes to the lowered fitness equals p/q. If the frequency of the recessive allele, *a*, is low, the onus for producing ill-adapted homozygotes rests almost entirely with the heterozygotes.

The role of heterozygotes in producing homozygous recessives can be illustrated verbally as well as by diagram. The mating of heterozygotes gives rise to homozygous recessives: $2pq \times 2pq \times 0.25 = p^2q^2$. The mating of homozygotes (*aa* × *aa*) also gives rise to homozygous individuals: $q^2 \times q^2 \times 1.00 = q^4$. The mating of heterozygotes with homozygous recessives also produces *aa* individuals; in this case, the blame rests equally with each type of parent. This blame can be calculated as $0.5 \times 2 \times 2pq \times q^2 \times 0.5$, or pq^3 for each. The ratio of responsibilities of heterozygotes and homozygotes for the production of ill-adapted homozygotes equals $(p^2q^2 + pq^3)/(pq^3 + q^4)$, or

p/q. Thus, to repeat an earlier conclusion, if q is small, heterozygotes must bear the onus for lowering the average fitness of a population by virtue of the aa individuals among their offspring.

W. D. Hamilton (1964a, b) introduced the concept of *inclusive fitness*, which can be defined as follows: the inclusive fitness of an individual bearer of a given allele equals the fitness of that bearer plus those of all other individuals, of this or future generations, who bear the same allele, weighted in each case by the degree of relatedness to the designated bearer. Producing ill-adapted aa homozygotes lowers the heterozygotes' inclusive fitness. This matter will not be discussed further at this time except to note that the standard assignment of fitness values is not necessarily the only possible one, but its use greatly simplifies subsequent calculations. These simplified calculations will be used in the subsequent examples.

Although Haldane (1937) first drew attention to the relationship between loss of fitness and gene mutation, I shall illustrate his results by using a computation similar to one employed by Crow (1948). Two items are to be understood. First, A is said to mutate to a at a rate u per generation. In the course of one generation, the frequency of A declines from p to $p(1-u)$ and the frequency of a increases correspondingly from q to $(q+up)$. Second, if the ratio p_1/q_1 in one generation equals the ratio p_2/q_2 in the next, then $p_1 = p_2 = \hat{p}$, where the "hat" ($\hat{\ }$) designates an equilibrium value.

In the case of a harmful recessive allele such as that illustrated above (fitnesses of AA, Aa, and aa individuals equal 1, 1, and $1-s$), the following calculations can be made

$$\frac{p}{q} = \frac{(p-up)}{(q+up-sq^2)}.$$

The solution to this equation is

$$sq^2 = u.$$

Now, if the average fitness ($\overline{W}$) of a population equals $1 - sq^2$, then it also equals $1 - u$. The mutation of a "normal" or "wild-type" allele (A) to a selectively deleterious, recessive allele (a) lowers the average fitness of a population by an amount equal to mutation rate (u)—an amount that is independent of the degree (s) to which aa individuals are selectively inferior.

If the allele a were not completely recessive but, when in heterozygous condition, expressed a fraction (h) of its deleterious effect on homozygotes (s), the following calculations are possible:

Genotype	AA	Aa	aa
Frequency	p^2	$2pq$	q^2
Fitness	1	$1 - hs$	$1 - s$
Frequency after selection	p^2	$2pq - 2hspq$	$q^2 - sq^2$
Average fitness		$1 - 2hspq - sq^2$	
Or, approximately		$1 - 2hsq$	

In estimating the effect of a partially dominant gene on the average fitness of a population, we can proceed as follows:

$$\frac{p}{q} = \frac{(p - up)}{[q + up - \frac{1}{2}(2pq)hs]} \text{ (approximately).}$$

Therefore, $hsq = u$ (approximately).

Because $\overline{W}$ in this instance equals $1 - 2hsq$, it also equals $1 - 2u$. The mutation of a wild-type allele (A) to a partially dominant (or incompletely recessive) allele (a) lowers the average fitness of a population by an amount equal to twice the mutation rate. This amount is independent of both s and h, measures of the harm done by the mutant allele to homozygous (s) and heterozygous (hs) carriers.

Calculations of this sort can be extended to sex-linked mutations as well. If individuals of the two sexes (XX, females; XY, males) are equally frequent, one-third of all X chromosomes are carried by males. Because deleterious sex-linked alleles are exposed to the action of selection in males (during every third generation, on the average), they do not accumulate to high frequencies. The equilibrium frequency of an allele (a) that reduces the fitness of homozygous (aa) females and of hemizygous (aY) males by an amount s equals $3u/s$ because at this frequency $(\frac{1}{3})(3u/s)(s) = u$, the rate at which the mutant allele arises by mutation. The amount by which the average fitness of males is lowered by the allele *a* equals $(3u/s)(s)$, or $3u$. Males, however, constitute only one-half of the population; consequently, the average fitness of the population is lowered by $1.5u$. Again, the degree of harm caused by the allele *a* to its individual carriers does not affect the extent by which the average fitness of a population is lowered as the result of gene mutation. The effect of variation on fitness is to lower it by amounts that are simple multiples (1, 2, 1.5) of the mutation rate itself.

Haldane's conclusions were also derived by Crow (1948) in a paper dealing with the yield of agricultural crops (Crow equated yield with fitness) and by Muller (1950) in what has become a classic paper, "Our Load of Mutations." The term *genetic load* derives from Muller's account of the dysgenic effect of mutation on human populations; in his account, however, the term referred both to the statistical consequences of mutation and to the debilitating effects of the mutations on their carriers. These two aspects of genetic load were separated by Crow (1958), who *defined* genetic load as the proportional decrease in average fitness (or other measurable quantity) of a population relative to the genotype possessing the maximum or optimum value. In symbols,

$$\text{Genetic load} = L = \frac{(W_{max} - \overline{W})}{W_{max}}.$$

Following the convention that maximum fitness (W_{max}) be assigned the value 1.00, this equation becomes

$$L = 1 - \overline{W}.$$

The term *load* (L) quickly supplanted Haldane's term *effect*; the decreases of average fitness by amounts equal to u, $2u$, and $1.5u$ became the genetic loads of populations.

The more or less immediate effects that "Our Load of Mutations" had upon population genetics and population geneticists can be organized under a series of headings: the classification of loads, estimating mutation rates from loads, estimating loads from mutation rates, and determining the effect of altered mutation rates on estimations of fitness.

The classification of loads

By definition, a genetic load is to be found in any population that possesses genetic variation with respect to fitness. Individual fitness values must fall both above and below their mean; therefore, the mean fitness of a population must be lower than that of the optimal genotype. Many geneticists felt that the sources of variation should be identified and that the loads to which they gave rise should be appropriately identified and labeled. Thus, the load identified by Haldane (1937), Crow (1948), and Muller (1950) became known as the *mutational load*; it was caused by recurrent mutation.

Early in the development of population genetics (Fisher, 1930), it was realized that both alleles (A and a) would be retained at stable equilibrium frequencies in a population if heterozygous (Aa) individuals possessed the highest average fitness. R. K. Nabours and L. L. Kingsley (1934) used standard Mendelian crosses to demonstrate such heterosis for a lethal factor in the locust *Apotettix eurycephalus*; the proportions of lethal heterozygotes obtained in F_2 and backcrosses consistently exceeded the expected ones. Again, starting in the mid-1940s, Wright and Dobzhansky (S. Wright and Dobzhansky, 1946; Dobzhansky, 1947, 1948, 1950) showed that inversion heterozygotes in laboratory populations of *Drosophila pseudoobscura* generally exceeded inversion homozygotes in fitness. The variation in fitness that characterizes a population within which heterozygous individuals possess maximum fitness leads to a genetic load: the *segregational load*, or, alternatively, *balanced load*.

A number of cases are known in which mothers and their (unborn) children exhibit detrimental interactions. The Rh blood group provides the best-known example. Mothers who are *rh/rh* and whose husbands are either *Rh/Rh* or *Rh/rh* can become pregnant with an *Rh/rh* child. Fetal blood that enters the mother's circulation through a flaw in the placenta induces the production of anti-Rh antibodies in the mother. These antibodies, if they pass into that or any subsequent *Rh/rh* fetus, tend to destroy the fetus's red blood cells. The resulting condition, erythroblastosis foetalis, may require that a total blood transfusion be given the affected baby at birth. In emphasizing that the Rh factor is merely one of many possible ones, Haldane once recommended that a husband never provide blood for transfusing his wife. Interactions of these sorts, influencing selection coefficients as they do, create mother-child *incompatibility loads*.

If the average fitness of a population possesses a stable maximum value with both alleles (A and a) present, as in the cases of heterozygous advantage or of frequency-dependent selection (the relative fitness of each genotype decreases as its frequency increases), chance events may displace gene frequencies from their optimum values; the resulting decrease in average fitness has been called the *drift load*. Furthermore, the equilibrium frequencies obtained under natural selection in the case of frequency-dependent selection may result in an average fitness lower than that which is theoretically possible. Kimura (1983) has called this the *dysmetric load* of the population.

Under the assumption that geographically isolated populations adapt rapidly to their local environments, migrant individuals arriving from elsewhere can be said to introduce maladapted alleles into the recipient population. The resultant lowering of fitness can be regarded as a *migration load*.

The substitution of one allele for another in a population as a consequence of natural selection has been shown by Haldane (1957, 1960) to entail a "cost" to the population by virtue of the deaths of ill-adapted individuals. A discussion of the estimated cost of evolution will appear in a later chapter; for the moment, it is sufficient to state that Haldane's cost of evolution can be viewed as a *substitution load*.

The above account suffices to emphasize the multitude of factors that introduce genetic variation into populations and, hence, that can be identified as sources of genetic loads. The list given here cannot be complete because the possible sources of genetic loads are limited in number only by the imagination of those interested in naming such loads.

Before we leave this topic, the existence of nongenetic loads should be mentioned. Phenotypic variation among the individual members of a population gives rise to a phenotypic load that is analogous in all respects to the genetic load that has been discussed above. The phenotypic load may or may not have a genetic basis. Among the most important nongenetic sources of phenotypic variation are *age* and differing *microenvironments*. The protracted sub-Saharan famine of

Variation, Fitness, and Genetic Load

the 1980s showed television viewers throughout the world that the very young and the (relatively) very old are the first victims of environmental and physical stress. From biblical instruction, many persons are familiar with the contrasting fates of seeds that fall on stony ground and on fertile soil. The matter of phenotypic loads will recur in a later chapter; they are mentioned here merely to complete our account of loads, genetic or otherwise.

Estimating total mutation rates from loads

With the realization that the genetic (i.e., mutational) load equals mutation rate (or is a simple function of that rate) and is independent of the effect of each mutant allele on the fitness of its carriers, some persons saw an opportunity to measure total human mutation rate by simply scoring birth defects. Several assumptions underlying this type of research should be made explicit. First, the population is assumed to be at equilibrium. This is a common assumption in population genetics; otherwise, many simple (and mathematically useful) relationships disappear. Second, the relationship between mutational load and mutation rate at a single locus is assumed to hold as well for all loci combined: Letting $\bar{u}$ equal the mutation rate averaged over all gene loci, and n equal the total number of these loci, one can calculate the total mutational load as $1 - (1 - \bar{u})^n$, or (if n is much smaller than $1/u$) approximately $n\bar{u}$. Finally, one assumes that all defects can be seen and scored. Unfortunately, an unknown proportion of fertilized ova fail to implant, thus making their presence known to neither the potential mother nor her physician; this is the greatest weakness of this type of research.

Dobzhansky (1970) presented a convenient summary of the studies carried out on the incidence of birth defects among newborn babies (Table 2-1). Of the half million babies examined, 4.50% suffered from a detectable defect, one that presumably lowered viability to some extent. The observed frequency of abnormal births (4.50%) can be taken as an estimate of the genetic load, and thus the total mutation rate.

Table 2-1. The proportions of all human births in which the newborn baby suffered an observable defect. (After Dobzhansky, 1970.)

Source	No. of births	% Malformed
Britain	170,224	2.88
Europe (except Germany)	78,610	2.96
Germany	8,516	2.20
United States	144,769	8.76
Other countries	121,264	2.85
TOTAL	523,383	4.50

Estimating loads from mutation rates

A research strategy that is the opposite of the one described in the previous section involved an attempt to estimate total genetic load by measuring total mutation rate. The outstanding study of this sort was carried out during the 1950s at the Oak Ridge National Laboratory under the direction of William Russell in collaboration with Lee Russell, a well-known developmental geneticist.

The Russells and others interested in this approach to estimating genetic load recognized one extremely difficult technical problem: the contribution of mutation to a population's genetic load depends upon u but is independent of s, the degree of harm done to individual carriers of mutant alleles; however, the smaller the effect of a mutant allele, the more difficult it is to detect its origin by mutation. One can only guess whether mutation rates for minor viability modifiers are less than, equal to, or greater than those for lethal and other easily detected mutations (see Neel and Schull, 1956:205ff.).

The Russells concentrated on seven loci in the house mouse (*Mus musculus*) at which mutant alleles affecting observable aspects of the phenotype (and which did not interfere with one another's expression) were known. Mutations that have occurred at any one of the seven loci in the wild-type individual can be detected by mating control and irradiated homozygous, wild-type individuals with the seven-mutant, homozygous individuals of the opposite sex. The spontaneous mutation rates, averaged over all seven loci and multiplied by the (esti-

Variation, Fitness, and Genetic Load

Table 2-2. Spontaneous mutations at seven loci in the house mouse, *Mus musculus*. (Data from Russell, 1962, after U.N. Report, 1962:106.)

Spermatogonia	
Number of offspring	544,897
Number of mutations	32
Mutation/locus/gamete	0.84×10^{-5}
Oocytes	
Number of offspring	98,828
Number of mutations	1
Mutation/locus/gamete	0.14×10^{-5}

mated) total number of gene loci, gives the mutational load in mouse (and, by extrapolation, in human) populations.

A summary of the Russells' data is presented in Table 2-2. The spontaneous mutation rates of the two sexes clearly differ. One can, however, consider the average of the two rates (0.50×10^{-5}) as the mutation rate for calculating mutational load because the sexes contribute equally to their offspring. If the number of gene loci in the mouse equals 100,000, the mutational load would be 0.39; if the number of loci is only 10,000, the load would be 0.05. The latter value is close to the one arrived at by studying human birth defects. Because the total number of gene loci is not known for any higher organism, data of the sort discussed here do not permit precise interpretation. (M. J. Simmons and Crow [1977] have reviewed data on fitness mutations in *Drosophila*; they suggest that the mutational load may be as high as 50%.)

Determining the effect of altered mutation rates on estimates of fitness

Chapter 3 summarizes several studies of irradiated populations of *Drosophila melanogaster* in which an effort was made to determine experimentally the populations' fitnesses and to relate these fitnesses to the radiation histories of the populations. The studies are mentioned here only to complete the account of research programs that were developed by different persons in response to the mathematically established relationships between mutation rate and genetic load:

Fifty Years of Genetic Load

- $L = U$ in the case of recessive mutant alleles ($U \approx n\bar{u}$),
- $L = 2U$ in the case of incompletely recessive (= partially dominant) mutant alleles, and
- $L = 1.5U$ in the case of sex-linked mutant alleles (U, in this case, refers to sex-linked mutations only).

For some persons, the relationships were an invitation to classify genetic loads. For others who assumed that the calculated relationships were true, two courses were open: study the load in order to learn about total mutation rate, and estimate total mutation rate in order to estimate genetic load. In my own studies, the basic relationship itself was investigated.

Personal comments

The matter of exposing persons to unnecessary radiation (diagnostic and therapeutic X radiation, industrial radiation, radiation from isotopes used in research laboratories, and radioactive fallout from weapons tests) was during the 1950s (and still is) of great concern to biologists, especially geneticists. The U.S. National Academy of Sciences (1956) convened a special committee of geneticists to study the biological effects of radiation and to make recommendations regarding the maximum exposure that should be permissible under law.

As is often the case in such situations, the scientists on the committee demurred until reminded by their chairman, Warren Weaver, that decisions would, in fact, soon be made; the only question was whether these decisions would be made by persons (politicians and administrators) who knew nothing about radiation biology or by those on the committee who, although feeling that they knew too little, did know a great deal about radiation and its effects. Weaver's argument carried the day.

M. Demerec and B. P. Kaufman of the Department of Genetics, Carnegie Institution of Washington, were members of this special committee. At one time Dr. Demerec asked both me and Hermann Moser (a population geneticist of considerable ability; see, for example, Moser, 1958) to compute the effect on human populations of

Variation, Fitness, and Genetic Load

radiation exposure using certain background information and, from our results, to recommend a maximum permissible exposure. I begged off, arguing that all persons using the same background information would arrive at the same conclusion (although not necessarily at the same recommendation). I offered to try a different approach: In a population of 200 million persons, I reasoned, there are 400 million copies of each gene. If the mutation rate for recessive lethal alleles at each locus were 10^{-6}, the equilibrium frequency of these lethals would be 10^{-3}. That is, 400 new lethals should arise each generation, and the total number of lethal alleles at a given locus should be 400,000.

Let us argue that in exposing persons to radiation, we of the present generation should not induce more lethals than would happen by chance only once in ten generations; such a proposal assumes that future generations may be both more intelligent and less ignorant than we, and that our responsibility today is merely to avoid doing irreparable harm to the population during our temporary stewardship.

Under the Poisson distribution the mean and variances are equal; therefore, the variance in the number of lethal alleles per locus from generation to generation (as well as from locus to locus or even from population to population) equals 400,000, as well; the standard deviation equals (approximately) 630. Ten percent of a (near) normal distribution lies beyond 1.3 standard deviations; hence, no more than 420 new mutations should be induced by man-made radiation (to be added to the 400 of spontaneous origin) in our generation. Because this number is nearly identical with the number arising spontaneously, a "doubling" dose of radiation should, by this reasoning, be the maximum permitted. Estimates of the doubling dose range from 3 R to 150 R (Dubinin, 1964).

I reported to Dr. Demerec after several days that a maximum permissible dose as low as 3 R could be defended. He reflected for a moment. "No," he finally said, "we have already decided on 10 Roentgens; that's the exposure that must be justified."

Questions concerning the harm done to human populations exposed to radiation remain unresolved despite the deliberations of Weaver's committee and those of the United Nations Scientific Committee on the Effects of Atomic Radiation (1962). While follow-up studies of the Japanese (and their later-born children) who were exposed to radiation during the bombings of Hiroshima and Nagasaki have revealed less evidence of genetic damage than expected, current accounts of persons affected by the Chernobyl nuclear power plant accident of 1986 suggest that the harm done to exposed individuals exceeds that which had been anticipated.

3

STUDIES OF IRRADIATED POPULATIONS

The effect of autosomal gene mutations on the average fitness of a population, according to calculations presented in the previous chapter, is to lower that average by an amount equal to (or two times in the case of partial dominance) the mutation rate itself; this effect is independent of the effect of this or that mutant gene on the fitness of individual homozygous or heterozygous carriers: the calculated effect of mutations at equilibrium, that is, involves u (mutation rate) but neither h (degree of dominance) nor s (detriment to homozygous carriers).

The term *average fitness of a population* requires a brief comment. The fitnesses under discussion are Darwinian fitnesses that measure the *relative* abilities of individuals of different genotypes to survive and reproduce within the populations in which they are found. To think that the average fitness ($\overline{W}$) of a population (which is an average of relative fitnesses of the members of that population) can be compared with the average fitness (also $\overline{W}$) of a different population can be quite misleading. Interpopulation comparisons should be made on the basis of a standard measure against which the individual members of *both* populations are measured. The average heights of Britons and Yugoslavs can be compared in terms of feet or meters; averages based on relative heights within populations (where, for example, the maximum height in each is said to be 1.00) would be useless in making an interpopulation comparison. Despite this obvious caveat, $\overline{W}$'s of different populations are often compared, even though the maximum fitnesses of the two (each of which is assigned a fitness of 1.00) have not (and, perhaps, cannot) be compared. Population fitness, a measure that enables one to make *inter*population comparisons, is an extremely elusive concept that will recur only later in this book.

The objective of the experiments summarized in this chapter (more extensive summary accounts as well as references to individual research papers can be found in Wallace, 1968a and 1981) was stated in the previous chapter: to confirm, if possible, the postulated relationship between mutational loads and rates of mutation. Before the experiments were undertaken I had little doubt that the calculations would be confirmed; one of the questions uppermost in my mind at that time was, What direction should my research take after these studies have been completed?

During July 1949 and April 1950 laboratory populations of *Drosophila melanogaster* (the vinegar fly) were set up; five of these were retained for intensive analysis (see Table 3-1). At the time, with a callousness that would be outrageous today, population 1 was referred to as the "atomic bomb" population; it was exposed, as the table shows, to a single heavy dose of X radiation. Population 7 was retained for study not only because it promised to yield results that would differ clearly from those of the control and more heavily irradiated populations but also because some radiologists apparently believed that 300 R per generation was the maximum exposure that any species could tolerate on a continuing basis. Populations 5 and 6 were included to reveal the role of population size in determining the fate of irradiated populations (a role that conceivably included the number of generations before an irradiated population became extinct).

Analytical procedures: *ClB*-like techniques

The results obtained by studying the irradiated populations are interesting, of course, and will be discussed directly. First, however, the experimental procedures themselves should be discussed at length. A discussion of techniques is made all the more important by the lapse of one or more scientific generations since the original studies were car-

Studies of Irradiated Populations

Table 3-1. Short histories of five experimental populations of *Drosophila melanogaster* studied over a period of five or six years at Cold Spring Harbor, N.Y., in an effort to learn of the effect of radiation exposure on a population's fitness. (Wallace, 1952.)

Population	Approximate size (no. adults)	Starting date	First sample no.	Irradiation
1	10,000	7/25/1949	1	Original ♂♂ 7000-R X ray Original ♀♀ 1000-R X ray
3	10,000	7/25/1949	1	None
5	1000	4/1/1950	20	2000 R/gen., chronic, gamma rays
6	10,000	4/15/1950	21	2000 R/gen., chronic, gamma rays
7	10,000	4/15/1950	21	300 R/gen., chronic, gamma rays

ried out. With the advent of electrophoretic techniques that permit the isolation and chemical analyses of individual allozyme variants, and of recombinant DNA techniques that permit the "reading" of base-pair sequences in DNA with relatively little effort, the cruder techniques of the past have been largely abandoned (Thompson, 1985). Few of those who study molecular evolution have had personal experience with the *ClB*-like procedures used by earlier *Drosophila* workers. Whereas these earlier workers, through intimate experience in manipulating chromosomes, were familiar with the idiosyncrasies, shortcomings, and underlying assumptions of these chromosomal analyses, younger workers are likely to know neither the procedures nor the uses to which they have been put. I am not embarrassed by this review of ancient history because I have come to realize that when molecular biologists eventually attempt to relate their observations to fitness (and evolutionary changes in populations), they will, of necessity, need to study the relative viabilities and reproductive success of individuals possessing contrasting genotypes. Energetics, oxygen consumption, and similar physiological studies are not conclusive in themselves; the stamina of the mule, for example, does not compensate for its sterility: despite its (physical) hybrid vigor, the mule's *euheterosis* (heterosis with respect to fitness [Dobzhansky, 1952]) is nil.

The crosses

The main features of *ClB*-like procedures are the use of (1) chromosomal inversions to suppress gene recombination in female *Drosophila* (gene recombination does not occur in males of most *Drosophila* species), and (2) dominant visible mutations that reveal which flies do and which ones do not carry the genetically altered chromosomes. Many dominant visible mutations are lethal when homozygous. If this is true for two mutants (say, M_1 and M_2), true-breeding, balanced-lethal strains (M_1/M_2) can be established and retained for supplying the large number of M_1/M_2 virgin females that are needed for the analyses of populations.

Inversion-containing chromosomes genetically marked with *Cy* (curly wings) and *L* (lobed eyes) or with *Pm* (plum eye color) facilitate the study of the second chromosome (one of the two large V-shaped autosomes) of *Drosophila melanogaster*. Crosses that lead to two different sorts of F_3 test cultures are illustrated in Figure 3-1.

Homozygous tests

Wild-type males from a population are mated individually with *Cy L/Pm* females to assay the *sorts* of chromosomes that are present within the population. A single F_1 male (which, of necessity, must carry one or the other of the two wild-type second chromosomes of the parental male) is crossed with *Cy L/Pm* virgin females. *Cy L/+* F_2 males and virgin females, five to seven of each, are then mated. Among the F_3 flies one expects to find *Cy L/+* and *+/+* flies in a ratio of two to one; that is, $33\frac{1}{3}\%$ of the F_3 flies are expected to be wild-type (recall that *Cy L/Cy L* flies die because *Cy* is a recessive lethal.)

The wild-type flies in these cultures are homozygous for one of the two chromosomes carried by the original wild-type male. If this chromosome carries a lethal allele at any locus (or, less likely, a multi-locus combination of alleles that is lethal when homozygous; see Thompson, 1986, for a recent review), the number of wild-type flies will be zero in that test culture. (By convention, chromosomes that lower the proportion of their homozygous carriers to 10% or less of the expected proportion are called *lethals*.) Among homozygous test cultures one often finds a bimodal distribution of viabilities: many of

Studies of Irradiated Populations

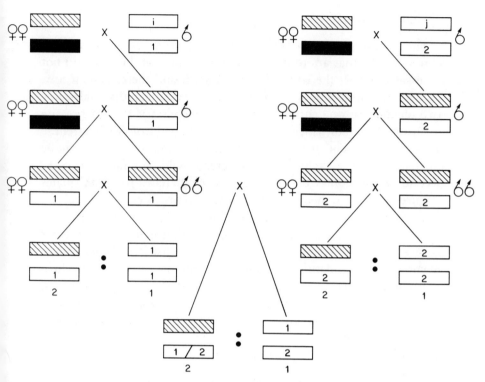

Figure 3-1. A mating scheme (the Cy L test) in which wild-type chromosomes carried initially by male *Drosophila melanogaster* in a freely breeding population are manipulated so as to produce individuals that are homozygous (1/1 and 2/2) for given chromosomes or heterozygous (1/2) for chromosomes of different origin. Cross-hatched bars represent genetically marked *Curly-Lobe* (*Cy L*) chromosomes; solid black bars represent *Plum* (*Pm*) chromosomes; and the open bars (labeled *i, j, 1,* and 2) represent various wild-type chromosomes. The expected frequencies of *Cy L*/+ and wild-type flies in the F_3 test cultures are 66.7% and 33.3%; distortions of these expected frequencies reflect differences in the egg-to-adult viability and in the speed of development of *Curly-Lobe* and wild-type flies.

the chromosomes occurring within a population are lethal or near lethal to their homozygous carriers; the remainder are distributed about—but, on the average, slightly below—the expected frequency ($33\frac{1}{3}\%$). Relatively few chromosomes when homozygous result in intermediate viabilities (10–20% of the expected frequency, for example).

Heterozygous tests

The homozygous tests of the preceding section reveal the sorts of *chromosomes* that are to be found in a population, but they do not necessarily reveal the sorts of *individuals* found there. Few, if any, individuals within a population of randomly mating individuals carry two identical second chromosomes; on the contrary, they carry pairs of chromosomes of which one has been obtained from each parent.

The crossing scheme used to create homozygous individuals can be used, with only a slight change, to create wild-type flies that carry second chromosomes derived from two different "parental" males (Figure 3-1). These heterozygous combinations of chromosomes are generated by crosses that are set up before the experimenter knows the nature of the two chromosomes involved; that is, the experimenter generates random combinations in the same sense that the individual members of a population carry random combinations of second chromosomes. Furthermore, by the use of systematic crossing procedures, the sorts of chromosomes making up each heterozygous F_3 combination eventually become known. In a study of n chromosomes, the n homozygous and n heterozygous test cultures can be listed as follows: *1/1, 1/2, 2/2, 2/3, 3/3, . . . , n/n,* and *n/1*.

Interpretation of the data (general)

What is the meaning of the data yielded by the *Cy L* test described above, and to what use can the data be put? Before considering the finer details, the following general comments can be made:

- The zygotes in the F_3 generation are assumed to be produced in Mendelian proportions: 1 *Cy L/Cy L* : 2 *Cy L/+* : 1 *+/+*.
- The number of eggs laid by five to seven F_2 females exceeds the number of progeny that a half-pint culture bottle can support without stress; some larvae are drowned before or just after hatching from the eggs, others starve, and still others develop at a reduced rate. These fates do not affect *Cy L/+* and *+/+* flies equally.
- Distortions from the expected ratio of two *Cy L/+* to one *+/+* are said to reflect the relative egg-to-adult viabilities of the genetically marked and wild-type flies.

- Egg-to-adult viability is considered to be a major component of Darwinian fitness.
- With reference to genetic load calculations (where L is said to equal a simple multiple of U), the expected relationship is not negated by (1) studying only a portion of the genome, (2) restricting measurements to a single fitness component (egg-to-adult viability), or (3) analyzing preequilibrium populations. The expectation that high mutation rate should correspond in some manner to high genetic load (low fitness) remains unchanged.

Earlier in this chapter I made the point that the average fitness of a population is an average of *relative* fitnesses found within that population and, hence, that average fitnesses of different populations cannot be safely compared. Under the general considerations of this section, then, the meaning of the experimental results should be made clear. First, the magnitude of the data should be appreciated: for the five best-studied populations, approximately 35,000 test cultures (or 11,000,000 flies) were studied; about half (17,500) of these cultures represented homozygous tests, and the remainder were heterozygous ones. Each of these heterozygous chromosomal combinations could have occurred as well within its corresponding population. The egg-to-adult viability of each heterozygous combination has been evaluated relative to that of $Cy\ L/+$ flies. Consequently, the *average* relative viability calculated for each population is an average based on numerous combinations of chromosomes. Furthermore, because the standard for comparison ($Cy\ L/+$ flies) is, as a close approximation, constant from population to population, the averages of the relative viabilities can, in this case, be used for interpopulation comparisons.

Interpretation of the data (specific)

Persons who have worked with *ClB*-like procedures such as the $Cy\ L$ tests described here frequently use the terms *frequency of wild-type flies, viability of wild-type flies,* and *relative fitness of wild-type flies* interchangeably. This is not strictly proper, especially for homozygous tests. Consider, for example, a test culture containing twice as many young $Cy\ L/+$ zygotes as $+/+$ ones. One can imagine that the $+/+$

homozygotes have an egg-to-adult viability relative to that of $Cy\ L/+$ zygotes equal to $1 - s$, where s may vary from 0 to 1.00. Further, one can imagine that the disadvantage (s) that characterizes the wild-type homozygotes is expressed in $Cy\ L/+$ individuals to an extent h (h, then, is a measure of the dominance of the wild-type chromosome; it can vary from 0 to 1.00). The percentages of wild-type flies expected in cultures where the wild-type chromosomes possess various combinations of s and h are shown in Figure 3-2. If the wild-type chromosome is recessive ($h = 0$), the proportion of wild-type homozygotes is not linearly related to s, but, nevertheless, as s increases, the proportion of wild-type flies decreases.

The effect of partial dominance is quite different: as h increases (holding s constant), so does the proportion of wild-type homozygotes. ClB-like test procedures cannot detect dominant viability mutations. Much of Figure 3-2 is of little practical concern. The lower figure in each cell of the table is the reciprocal of the harm done to heterozygous individuals (equal to $1/hs$); this number equals the average number of generations that a mutation with the specified effect on its heterozygous carriers would persist in a population. Dominant lethals (lower right corner) kill their carriers as early zygotes, so they are not found in populations. Much of the remainder of the table concerns mutations that would persist in populations for fewer than ten generations. The portions of the table of greatest practical interest are the extreme left and top margins.

In the case of heterozygous combinations, each of the two wild-type chromosomes is characterized by its own s (s_1 and s_2); consequently, an array (Figure 3-3) illustrating only two degrees of dominance ($h = 0.10, h = 0.20$) has been compiled. The same degree of dominance has been assigned to each of the two chromosomes. In Figure 3-3 it is clear that as s increases, the proportion of wild-type flies ($+_1/+_2$) in the test cultures decreases. The same is true for h: as h increases, the proportion of wild-type heterozygotes decreases.

The absence of a strictly linear relation between the frequency of wild-type flies in the F_3 test cultures and the relative viability of these flies with respect to recessive mutations is only a nuisance; that dominance can cause an actual misinterpretation of homozygous tests is a matter of much more serious concern. This problem can be circumvented, however, by a slight modification of the crossing scheme (see

Studies of Irradiated Populations

s \ h	0	.1	.2	.3	.4	.5	.6	.7	.8	.9	1.0
0	.333	.333	.333	.333	.333	.333	.333	.333	.333	.333	.333
.1	.310	.313 100	.315 50	.317 33	.319 25	.321 20	.324 17	.326 14	.328 13	.331 11	.333 10
.2	.286	.290 50	.294 25	.299 17	.303 13	.308 10	.313 8	.317 7	.323 6	.328 6	.333 5
.3	.259	.265 33	.271 17	.278 11	.285 8	.292 7	.299 6	.307 5	.315 4	.324 4	.333 3
.4	.231	.238 25	.246 13	.254 8	.263 6	.273 5	.283 4	.294 4	.306 3	.319 3	.333 3
.5	.200	.208 20	.217 10	.227 7	.238 5	.250 4	.263 3	.278 3	.294 3	.313 2	.333 2
.6	.167	.175 17	.185 8	.196 6	.208 4	.222 3	.238 3	.256 2	.278 2	.303 2	.333 2
.7	.130	.139 14	.149 7	.160 5	.172 4	.188 3	.205 2	.227 2	.254 2	.288 2	.333 2
.8	.091	.098 13	.106 6	.117 4	.128 3	.143 3	.161 2	.185 2	.217 2	.263 1	.333 1
.9	.048	.052 11	.057 6	.064 4	.072 3	.083 2	.098 2	.119 2	.152 1	.208 1	.333 1
1.0	0	0 10	0 5	0 3	0 3	0 2	0 2	0 1	0 1	0 1	— 0

Figure 3-2. The frequencies of wild-type homozygous flies in the F_3 cultures of the Cy L test where the expected frequency of these flies is 0.333. The frequencies listed in the leftmost column, reading from top to bottom, are those expected of chromosomes carrying recessive mutations ($h = 0$) with increasingly severe deleterious effects on the viability of their carriers, culminating in complete lethality ($s = 1$). The topmost row merely confirms that if a chromosome exerts no deleterious effect on its carriers, dominance is of no consequence. The remainder of the table illustrates the frequencies of wild-type homozygotes given certain combinations of s (effect on homozygotes) and h (degree of dominance). For a given h (other than complete dominance), increasing levels of harm lead to decreasing proportions of wild-type flies, as expected. For a given s, on the other hand, increasing levels of dominance lead to *increasing* proportions of wild-type homozygotes because of the deleterious effect of the mutation-bearing chromosome on the $Cy L/+$ flies. The lower figure in each cell equals $1/hs$ and is a measure of the average number of generations such a mutant would persist in a population; most combinations shown here would persist for very short times (i.e., they would be found only rarely).

Figure 3-4): If F_2 $Pm/+$ males are used for mating rather than $Cy L/+$ males, test cultures containing four, rather than two, classes of flies are obtained. In these cultures, the $Cy L/Pm$ flies can be used as the standard for comparison. These flies do not carry the wild-type chromosome and, hence, are not affected by the dominance or lack of dominance of the mutant genes carried by that chromosome.

If the $Cy L-Pm$ (a four-class as opposed to the two-class $Cy L$)

Fifty Years of Genetic Load

h = 0.1 top; h = 0.2 bottom

s_2 \ s_1	0	.1	.2	.3	.4	.5	.6	.7	.8	.9	1.0
0	.333 .333	.332 .331	.331 .329	.330 .326	.329 .324	.328 .321	.326 .319	.325 .316	.324 .313	.323 .311	.321 .308
.1	.332 .331	.331 .329	.330 .326	.329 .324	.328 .321	.326 .319	.325 .316	.324 .313	.323 .311	.321 .308	.320 .305
.2	.331 .329	.330 .326	.329 .324	.328 .321	.326 .319	.325 .316	.324 .313	.323 .311	.321 .308	.320 .305	.319 .302
.3	.330 .326	.329 .324	.328 .321	.326 .319	.325 .316	.324 .313	.323 .311	.321 .308	.320 .305	.319 .302	.318 .298
.4	.329 .324	.328 .321	.326 .319	.325 .316	.324 .313	.323 .311	.321 .308	.320 .305	.319 .302	.318 .298	.316 .295
.5	.328 .321	.326 .319	.325 .316	.324 .313	.323 .311	.321 .308	.320 .305	.319 .302	.318 .298	.316 .295	.315 .292
.6	.326 .319	.325 .316	.324 .313	.323 .311	.321 .308	.320 .305	.319 .302	.318 .298	.316 .295	.315 .292	.313 .288
.7	.325 .316	.324 .313	.323 .311	.321 .308	.320 .305	.319 .302	.318 .298	.316 .295	.315 .292	.313 .288	.312 .284
.8	.324 .313	.323 .311	.321 .308	.320 .305	.319 .302	.318 .298	.316 .295	.315 .292	.313 .288	.312 .284	.311 .281
.9	.323 .311	.321 .308	.320 .305	.319 .302	.318 .298	.316 .295	.315 .292	.313 .288	.312 .284	.311 .281	.309 .277
1.0	.321 .308	.320 .305	.319 .302	.318 .298	.316 .295	.315 .292	.313 .288	.312 .284	.311 .281	.309 .277	.308 .273

Figure 3-3. The frequencies of wild-type heterozygous flies in the F_3 cultures of the $Cy\ L$ test where the expected frequency of these flies is 0.333. Along the left and top margins are listed from top to bottom and from left to right the increasingly deleterious effects (s_1 and s_2) of these chromosomes on their homozygous carriers. The top figure of the two listed in each cell of the table gives the frequency of wild-type flies expected if each chromosome has a dominance (h) of 0.10; the bottom figure gives the frequency if the dominance of each is 0.20. Unlike Figure 3-2, the expected frequencies of heterozygous wild-type flies decline both from right to left and from top to bottom in the table; h, in this tabulation, is manifest in the wild-type as well as in the $Cy\ L/+$ flies.

procedure removes such a worrisome problem concerning dominance, why has it not become the standard experimental procedure? First, because sorting and counting four instead of two classes of flies more than doubles the amount of labor required, thus substantially reducing the number of tests that can be performed; and, second, because the number of flies per test culture that provide information concerning the relative viabilities of wild-type flies is only half the number one obtains by the $Cy\ L$ procedure (two-class test cultures); hence, the statistical errors of these estimated viabilities are greatly increased. The prospect of performing fewer four-class tests, each of which would provide statistically less reliable information, has caused many workers to use the simpler $Cy\ L$ procedure.

Studies of Irradiated Populations

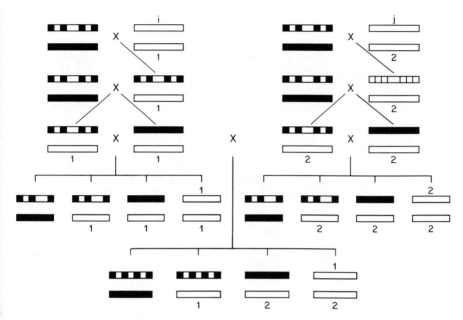

Figure 3-4. Mating scheme (the *Cy L–Pm* test) in which wild-type chromosomes carried initially by male flies in a freely breeding population are manipulated so as to produce F_3 individuals that are homozygous (1/1 and 2/2) for given chromosomes or heterozygous (1/2) for chromosomes of different origin. Black-and-white bars represent the genetically marked *Cy L* chromosomes; solid black bars represent *Pm* chromosomes; and the open bars labeled *i, j, 1,* and *2* represent wild-type chromosomes. The use of *Pm*/+ males (as opposed to the *Cy L*/+ males of Figure 3-1) in the final cross leads to "four-class" progeny in which *Cy L/Pm*, *Cy L*/+, *Pm*/+, and +/+ flies are expected in equal proportions. The observed proportions, of course, reflect differences in the viability of the different types of flies. Use of *Cy L/Pm* flies as a standard for comparison removes the dominance of the wild-type chromosome as a possible complicating factor in the interpretation of data of this sort.

My experience with *Cy L* and *Cy L–Pm* tests has led me to several pragmatic conclusions:

- Relative egg-to-adult viabilities that are revealed by *ClB*-like procedures are of little or no value in comparing geographically remote populations except with respect to the major sorts of chromosomes found in each: lethals, semilethals, and quasi-normals. That the average frequency of wild-type flies heterozygous for random combinations of chromosomes obtained from a local population in New York may be 0.3457 while the corresponding figure

for a Texan population is 0.3614 is, in my opinion, of no value in comparing these two widely separated populations; too many extraneous variables of possible importance remain unknown.
- Laboratory populations of *Drosophila* of relatively recent origin whose experimental treatments have been well documented and whose founder stocks were strictly comparable can be reliably compared by means of *ClB*-like procedures.
- Average effects based on large numbers of tests are reliable; effects based on the results of individual cultures may be atypical or idiosyncratic. Thus, the average frequency of F_3 wild-type heterozygotes carrying chromosomes from population 1 exceeded that of corresponding heterozygotes carrying chromosomes from population 3. This was true for cultures yielding only *Cy L*/+ and +/+ flies where systematic interactions (either dominance or homozygosis for common alleles, for example) between the *Cy L* and + chromosomes *could* have distorted the observed frequencies; the use of the four-class *Cy L–Pm* mating system ruled out this possibility.

A number of observations reveal that the results of individual test cultures (or tests of individual chromosomes) must be accepted with caution. At the extreme are instances in which a wild-type chromosome that produces viable *Cy L*/+ heterozygotes fails to produce viable *Pm*/+ flies. Presumably such a chromosome carries a lethal that is allelic to *Pm* or is allelic to another lethal present on the *Pm* but not the *Cy L* chromosome.

The interactions need not always be this striking, however. I showed that, on the average, wild-type chromosomes that resulted in M_1/+ flies of poor viability also resulted in M_2/+ flies of poor—but not *as* poor—viability; the inverse (reversing M_1 and M_2) was also true. On the other hand, wild-type chromosomes that resulted in M_1/+ flies of high viability resulted in M_2/+ flies of high—but not *as* high— viability; the inverse was also true in this case. (This study [Wallace, 1963a] was based on *D. pseudoobscura*; the dominant marker genes (*M*) were *Bare, Lobe, Blade Scute,* and *Delta.*)

The extent to which even homozygous viabilities measured by means of *ClB*-like procedures can be characteristic of particular genetic backgrounds (and concomitant interactions) has been demonstrated by Dobzhansky and B. Spassky (1944). Here, nine wild-type chromosomes, five of which produced homozygotes of poor and four of high viability, were tested in a series of "alien" genetic backgrounds representing five widely separated geographic localities. Without exception (see Table 3-2 and Figure 3-5) the chromosomes characterized by low viabilities in the original tests improved in the new genetic backgrounds; conversely, chromosomes initially exhibiting high viabilities exhibited lower ones in the new backgrounds (see Wallace, 1968a:147).

Because the following two points will be discussed again later in this chapter, they are mentioned here without further comment:

- The average viabilities observed for F_3 wild-type heterozygotes are correlated with the reproductive fitnesses of carriers of these same chromosomes in populations; this has been shown by comparing the average viabilities of wild-type carriers of sporadic lethals (presumably those found in populations because of recurrent mutation) with carriers of common lethals (those whose frequencies are so high for so many generations that one must postulate that they are retained in the population by natural selection).
- Non-Mendelian patterns of inheritance such as segregation distortion do not seem to have affected the frequencies of wild-type flies in the $Cy\ L$ test (F_3) cultures to any appreciable extent, if at all (Wallace and Blohowiak, 1985a).

Observations and their interpretation

An inordinate amount of time has been spent discussing the techniques by which the following (greatly condensed) data were acquired because such techniques are no longer familiar to many persons; hence the need to point out the danger of arriving at misleading interpretations. The data themselves can be presented in a single, simple table (Table 3-3). In brief, the data obtained by studying irradiated *Drosophila* populations did not agree with expectations based on the theoretical calculations relating mutation rate to genetic load. The disagreement between observation and expectation does not mean that the theoretical calculations are wrong; it means only that the

Table 3-2. The effect of background genotype on the viability of *Drosophila pseudoobscura* homozygous for various wild-type second and fourth chromosomes. The background genotypes consist of wild-type chromosomes obtained from the indicated geographic localities. (After Dobzhansky and Spassky, 1944.)

Chromosome	Original	Washington	Colorado	California	Mexico	Guatemala	Average
AA1003	34.6	32.1	33.9	33.2	28.7	29.9	31.5
AA1015	14.2	12.8	31.2	30.8	24.7	24.3	24.8
AA1178	34.8	35.0	31.0	33.6	31.9	32.1	32.7
KA667	33.6	34.0	29.9	32.9	29.8	33.4	32.0
KD745	24.4	33.6	31.6	32.7	29.9	32.3	32.0
AA955	32.8	28.4	31.8	31.1	30.5	28.5	30.1
AA1035	23.2	28.0	32.5	23.4	28.4	29.4	28.7
PA851	19.0	25.9	25.7	33.0	28.2	24.5	27.5
PA998	19.4	29.3	27.2	28.3	29.8	25.3	28.0
AVERAGE	26.2	28.8	30.5	31.0	29.1	28.9	29.7

Studies of Irradiated Populations

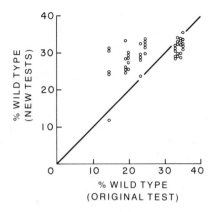

Figure 3-5. Comparison of the original viabilities exhibited by wild-type *Drosophila pseudoobscura* homozygous for one of nine different chromosomes (horizontal axis), and viabilities of these same homozygotes in each of five different genetic backgrounds (vertical axis). Note that the improvement of originally poor viabilities and the lessening of originally good ones are not the result of mere sampling errors (the number of flies counted in obtaining each point on the chart was too great to allow this explanation); rather, the altered relationships reflect magnitudes of gene interactions. (From Wallace, 1981, courtesy of Columbia University Press.)

calculations have neglected some aspect of the genetics of populations that should not have been ignored.

The unexpected results are presented in Table 3-3. First, however, it must be emphasized that the sorts of *chromosomes* in the various populations as revealed by homozygous tests did not differ from expectation. For example, population 1 contained more lethal and semilethal chromosomes as well as more chromosomes with detrimental effects on homozygous viability than did the control population (3); this state of affairs continued for at least two years following the initial (and only) exposure of population 1 to radiation. Similarly, under constant exposure to gamma radiation, populations 5, 6, and 7 accumulated lethal and semilethal chromosomes at faster rates than did the control population; the irradiated populations also had higher frequencies of minor deleterious chromosomes than did the control population. As would be expected, population 7 more nearly resembled the control population; the radiation exposure of population 7 was considerably lower than that of populations 5 and 6 (Table 3-1).

The sorts of *chromosomes* observed in the experimental popula-

37

Fifty Years of Genetic Load

Table 3-3. The average frequencies in F_3 test cultures of flies heterozygous for different wild-type second chromosomes from five experimental populations of Drosophila melanogaster. (Wallace, 1956.)

Population	Percent wild type		Relative fitness		No. of tests	
	1951	1954	1951	1954	1951	1954
1	35.02	34.80	1.04	1.03	994	3832
3*	33.75	33.75	1.00	1.00	722	3762
5	31.07	31.97	0.92	0.95	390	3176
6	31.95	33.17	0.95	0.98	707	3318
7	32.83	33.26	0.97	0.99	730	3391

*Control population.

tions may not have differed from those expected on the basis of their radiation histories, but the viabilities of flies carrying random combinations of these same chromosomes departed considerably from a priori expectations. Table 3-3 lists the summaries of early data (through 1951) and then the total data through 1954. The number of tests increased considerably between 1951 and 1954: from a range of 390–994 to 3176–3832 tests per population.

According to the earlier tests, the average viability of heterozygous combinations of chromosomes obtained from population 1 exceeded that of chromosomes obtained from population 3, despite the lethals, semilethals, and detrimental chromosomes that were characteristic of the irradiated population. The frequencies of wild-type heterozygous flies in tests of population 1 remained 3–4% higher (in relative terms) than those of the control population for a period of approximately 125 generations.

The other, continuously irradiated populations also failed to confirm expectations based on standard mathematical calculations. During the interval from 1951 to 1954 when lethals and other chromosomes with detrimental effects on their homozygous carriers increased in frequency, the viabilities of flies carrying heterozygous combinations of these same chromosomes steadily increased. Genetic load calculations predict that as deleterious mutations increase in frequency

Studies of Irradiated Populations

Table 3-4. Comparison of two measures of fitness to which the experimental populations of *Drosophila melanogaster* were subjected: the average frequency of flies heterozygous for wild-type chromosomes (see Table 3-3) and the average number of offspring produced by fertile females (single-pair matings). (After Wallace 1956, 1959a.)

		Measure based on the average no. of progeny per fertile pair	
Population	1954 F_3 tests	Generation I (no inbreeding)	Following ten generations of inbreeding
1	1.03	1.02	0.85
3	1.00	1.00	1.00
5	0.95	0.84	0.79
6	0.98	0.82	0.69
7	0.99	1.07	0.92

in a population, the average viability of their heterozygous carriers will decrease.

The unexpected relationship between populations 1 and 3 was rechecked using the more elaborate *Cy L–Pm* test illustrated in Figure 3-4. The higher average viability of the wild-type heterozygotes of population 1 was confirmed. Finally, a test of reproductive ability (progeny per female) was performed on flies from all populations (Table 3-4). Females hatching from eggs sampled directly from the different population cages produced progeny in numbers that paralleled in important ways the estimates of viability obtained by the *Cy L* and *Cy L–Pm* tests; females from population 7 (an irradiated population), however, produced the greatest number of progeny per female on the average. After ten generations of brother-sister matings, the situation with respect to progeny production changed drastically: all populations with a radiation history now produced substantially fewer progeny per female than the control. The low progeny production of the inbred females reflected once more the sorts of *chromosomes* to be found in these populations: not only did the chromosomes of irradiated populations lower egg-to-adult viability when homozygous, but they also reduced the ability of inbred (homozygous) females to lay eggs and produce viable offspring.

39

Fifty Years of Genetic Load

Recent reanalyses of the data on irradiated populations

The extensive data gathered from populations 1, 3, 5, 6, and 7, as well as from populations 17, 18, and 19, which were "daughter" populations of 5 and 6 (but removed from radiation exposure), have been recently reexamined (Wallace and Blohowiak, 1985a, b; Wallace, 1986a, b, 1987a) with a somewhat different purpose in mind. Recall that the main tests of the 1950s involved five populations from which egg samples were removed at frequent intervals over a period of several years; males hatching from these eggs initiated the crosses (Figures 3-1 and 3-4) that led eventually to F_3 cultures in which wild-type flies were either homozygous for individual chromosomes or heterozygous for random combinations of these chromosomes. As a rule, four technicians carried out these crosses and counted the flies that emerged in the F_3 cultures.

In the recent analyses, the frequencies of wild-type flies in both homozygous and heterozygous test cultures counted by one technician and representing one sample from one population have been normalized by dividing each percentage (culture by culture) by the average viability of the quasi-normal heterozygotes of that same technician's cultures and sample. The result has been the normalization of all samples of all technicians from all populations about a single mean viability, 1.00.

The new analysis has shown, by taking advantage of the enormous number of cultures that can now be consolidated for study, that segregation distortion (SD) is apparently not important in giving rise to an individual culture containing high frequencies of wild-type heterozygotes. This demonstration (Wallace and Blohowiak, 1985a) was possible because (1) segregation distortion affects males (in SD males, functional sperm carry one or the other of two homologous autosomes in high frequency rather than both in the expected 1:1 ratio), and (2) the systematically organized pattern of crosses allows one to identify which of the two chromosomes in any heterozygous cross came from the male parent and which from the female.

A second outcome has been the demonstration that frequencies of wild-type heterozygotes observed in test cultures are related (as had been assumed) to the fitness of similar heterozygotes in the popula-

Studies of Irradiated Populations

Table 3-5. Relative fitnesses of various genotypes in three experimental populations of *Drosophila melanogaster* within which abrupt and subsequently long-sustained increases in the frequency of lethal heterozygotes suggest a genetic "changeover." Nonlethal chromosomes, $+$; ordinary lethals, l_i or l_j; selectively favored lethals (those responsible for the increased probability of allelism), l_f. (Wallace, 1986a.)

Population	$+/+$	$+/l_i$	$+/l_f$	l_i/l_i	l_i/l_j	l_i/l_f	l_f/l_f
5	1.00	0.96	1.10	0	0.93	1.35	0
7	1.00	0.97	1.07	0	0.93	1.32	0
19	1.00	0.99	1.05	0	0.98	1.22	0

tions themselves. Certain lethals occurred in the experimental populations in exceptionally high frequencies; their presence was revealed by extended periods of exceptionally high frequencies of lethal heterozygous combinations of wild-type chromosomes. In populations 5, 6, and 7, nonlethal heterozygous combinations carrying one of these so-called common, or "favored," lethals produced higher percentages of wild-type flies than did corresponding combinations carrying a "sporadic," or "ordinary," lethal.

Of special importance is the estimation that has been made of the relative fitnesses of seven genotypes within the experimental populations themselves (Table 3-5). Here, $+$, l_i (or l_j), and l_f represent nonlethal, ordinary or sporadic lethal, and favored lethal chromosomes, respectively. The frequencies of these three types of chromosomes are known, as are the frequencies of allelism of lethals. This information allows the calculation of the relative fitnesses of $+/+$, $+/l_i$, $+/l_f$, l_i/l_i, l_i/l_j, l_i/l_f, and l_f/l_f flies where the fitness of $+/+$ flies equals 1.00 by definition. The fitness of l_i/l_i and l_f/l_f flies equals 0 because these flies die (they arc lethal homozygotes). In making these calculations it is also necessary to assume that the fitnesses of $+/l_i$ and l_i/l_j flies are $1 - s$ and $1 - 2s$, respectively (see Table 3-6).

These tests reveal that favored lethals, when heterozygous, can have fitnesses as much as 35% greater than the fitness of heterozygous flies carrying only nonlethal chromosomes (Wallace, 1986a, 1987a). These same analyses also reveal, as many persons have previously shown (Stern et al., 1952; Muller, 1950; Crow and Temin, 1964; Wallace, 1968a:172 ff.), that ordinary lethals tend to depress the viability of

Table 3-6. Viabilities of *Drosophila melanogaster* heterozygous for second chromosomes with drastic (lethal or semilethal) or nondrastic (quasi-normal) effects on the viabilities of their homozygous carriers. Viabilities are expressed here as percentage of wild-type flies in F_3 test cultures. (After Wallace and King, 1952.)

Population	2-Drastic	1-Drastic	0-Drastic
1	35.32	34.74	34.75
3	33.66	34.43	34.53
5	31.62	32.38	32.03
6	32.54	33.15	34.54
7	32.69	32.98	33.77
Average (5, 6, 7)	32.28	32.84	33.45
Decrease relative to control (0-drastic)	100%	75%	48%

their heterozygous carriers by 1% or more. The fitnesses discussed in this paragraph, it is important to note once again, are estimates of the average fitnesses of flies of the seven genotypes *within* the populations, themselves. They are *not* estimates based on distorted Mendelian ratios among flies raised in laboratory culture bottles.

Summary

The studies of irradiated *Drosophila* populations, studies that in the course of five to six years during the 1950s involved nearly 40,000 test cultures and the sorting and counting of well over 12,000,000 flies, were undertaken not to estimate total mutation rate from an empirically determined load or to estimate the mutational load from an analysis of mutation rates but, instead, to see whether the mathematically established relationship between mutation rate and genetic load could be demonstrated empirically. The experimental data did not support the simple algebraic model used by Haldane and others in carrying out their calculations. On the contrary, the results called for new assumptions, for a new model.

Two points emerge from the combined averages (5, 6, 7) in Table 3-6. First, carrying two drastic chromosomes has twice the detrimental effect of carrying only one (i.e., 32.84 is midway between 33.45 and 32.28). In brief, the fitnesses of $+/l_i$ and l_i/l_j are, it seems, $1 - s$ and $1 - 2s$. The second point may be noticed in passing: one-half of the difference between 34.53 (the viability of flies carrying two different quasi-normal chromosomes from the control population [3]) and 32.28 (the average viability of flies carrying two drastic chromosomes from the irradiated populations [5, 6, 7]) can be accounted for by the 0-drastic combinations of chromosomes from the irradiated populations.

A possible model

One possibility (one also proposed by Bonnier and his co-workers [1958]) can be visualized as follows:

Let the existing genetic variation at a locus be represented by A and a newly arisen variant by A^*. A^* may, of course, be lost from the population by accident no matter what its *average* effect on its heterozygous carriers might be. Otherwise, however, if A^* depresses the Darwinian fitness of its heterozygous carriers (and, initially, it will *only* occur in heterozygous individuals), it will tend to be eliminated more or less rapidly from the population as the result of natural selection. On the other hand, if A^* increases the Darwinian fitness of its heterozygous carriers, its frequency in the population will tend to increase, and to do so rapidly if the increase in fitness is substantial.

As the frequency of A^* increases, the fitness of A^*/A^* individuals becomes an important factor in determining its ultimate fate. If the fitness of these homozygotes exceeds that of A/A^* heterozygotes, the frequency of A^* will eventually reach 100% (see Nöthel, 1987). If the fitness of A^*/A^* individuals is lower than that of A/A^* heterozygotes (or even lower than that of A/A homozygotes), A^* will be retained in the population at an intermediate frequency; precisely what this frequency might be depends upon the relative fitnesses of the three genotypes: A/A, A/A^*, and A^*/A^*.

The possibility that heterozygous individuals might exceed both homozygotes in relative fitness was mentioned in the survey of genetic loads presented in Chapter 2. The algebraic treatment of this relationship takes a familiar form:

Fifty Years of Genetic Load

Genotype	A/A	A/A^*	A^*/A^*
Frequency	p^2	$2pq$	q^2
Fitness	$1-S$	1	$1-T$
Frequency after selection	$p^2 - Sp^2$	$2pq$	$q^2 - Tq^2$
Average fitness, $\overline{W}$	$1 - Sp^2 - Tq^2$		

If $p_0/p_1 = q_0/q_1$, then $p_0 = p_1 = \hat{p}$, and $q_0 = q_1 = \hat{q}$. By means of this relationship, one can show that $\hat{p} = T/(S + T)$, and $\hat{q} = S/(S + T)$. At equilibrium, $\overline{W}$ becomes $1 - ST/(S + T)$.

If $S = T = 1.00$ (that is, A/A and A^*/A^* individuals die before reproducing or are sterile), the average fitness of the population becomes 0.5; this corresponds to the situation that exists in a balanced-lethal strain of flies such as the *Cy L/Pm* strain that has been mentioned so often in this chapter: one-half of the progeny (the *Cy L/Pm* flies) produced in these stock bottles survives; the other half (the *Cy L/Cy L* and *Pm/Pm* homozygotes) dies.

The model proposed here to account for the observation that the relative fitness of a population need not necessarily reflect the sorts of chromosomes that are found in that population is one that can impose a tremendous genetic load (a segregational or balanced load) on a population. Any attempt, for example, to create a strain of organisms that carried many independent, balanced-lethal systems would be severely limited: the first system would permit only one-half of the progeny to survive; the second, one-fourth; the third, one-eighth; the fourth, one-sixteenth. In the case of mammals, in which there are many chromosomes as a rule and relatively few offspring, 10 balanced-lethal systems (a number that would allow only one surviving offspring per 1024 zygotes) would impose an intolerable burden on a population. In fact, during the 1950s one frequently heard discussions of "load space," the proportion of all zygotes that could be sacrificed in maintaining a genetic load without danger of the population's becoming extinct.

One avenue of escape from the problem is to imagine that individual gene loci are occupied by unsuspectedly large numbers of multiple alleles (Table 3-7). This solution, as we shall see in the following chapter, is not without its limitations as well. Limitations or not, however, the supposition that various heterozygotes can exhibit supe-

Table 3-7. The approximate average number of eggs that a female of a biparental species must produce in order to maintain balanced polymorphic systems involving a given number of independent gene loci and alleles which when homozygous have specified deleterious effects. Upper figures give the number of eggs required if only two alleles exist; lower ones, if twenty alleles interact heterotically. These calculations have been made according to the multiplicative model. (After Wallace, 1963b.)

Disadvantage of homozygotes	No. of loci			
	1	10	100	1000
1	4	2000	$>10^{30}$	$>10^{300}$
	3	3	340	$>10^{22}$
.5	3	36	$>10^{12}$	$>10^{124}$
	3	3	25	$>10^{10}$
.1	3	3	340	$>10^{22}$
	3	3	4	300
.01	3	3	4	300
	3	3	3	4

rior fitness finds concrete support in data such as those summarized in Table 3-5.

Returning once more to the situation that existed in the mid and late 1950s, we can note a problem other than that concerned with the magnitude of the genetic (segregational) load: If mutations that are selectively superior in heterozygous condition do, in fact, occur in populations, how often or at what rate do they occur? What proportion of all mutations enhances the fitness of their heterozygous carriers? One per hundred? One per thousand? One per million?

Laboratory populations of flies did not provide answers to the "new" questions. Timothy Prout (1954) showed that the effective sizes of the irradiated populations were not strikingly unlike the number of adult flies in the laboratory cages. This finding ruled out the possibility, for example, that all members of an experimental population, although numbering in the thousands, might be the progeny of 20 or fewer females. This was precisely the possibility that Gert Bonnier had attempted to rule out in his experiments (Bonnier et al., 1958). Granted, however, that 5000 adult females may lay as many as 300 eggs each (1,500,000 eggs in all), determining the frequency with which selectively favored mutational changes arise becomes impossible. Even favorable mutations that might arise only once in a million gametes would be encountered nearly every generation in such *Drosophila* populations. A new type of experimental approach was now needed, one that focused more precisely on the newer problems. The experimental populations had served their purpose; they were discontinued and the studies described in Chapter 4 were undertaken.

Personal comments

Linkage is a term that applies to scientific as well as diplomatic matters. Steven Jay Gould, because he has postulated rather novel patterns and mechanisms for evolution ("punctuated equilibria"; see Gould and Eldredge, 1977), has found himself linked to antievolutionists by creation "scientists." The latter seemingly conduct an endless search of the scientific literature looking for items that they believe will serve their cause—by sowing confusion if in no other way.

Lysenkoists, hard-pressed in their denial of genetics, once seized upon bacterial transformation as supporting their thesis that physiological, rather than genetic, mechanisms were responsible for heredity. DNA, the Lysenkoists claimed, was a component of physiological mechanisms (and, hence, anti-Mendelian) despite its discovery being the outcome of a long and diligent search by geneticists for the material basis of heredity.

In more recent years, belief in the asteroid theory of the Cretaceous-Tertiary extinction of dinosaurs has been linked to the "nuclear winter" model of climatic changes that could result from nuclear conflagrations. Skepticism concerning the asteroid theory is said by some proponents of that theory to reveal a willingness (or worse, a desire) on the part of their opponents to settle international disputes through nuclear warfare.

A similar linkage of views was practiced in the 1950s. To question the accepted view of the genetic structure of populations (that used by Haldane [1937], Muller [1950], and others, both earlier and later) was, in the eyes of the supporters of that view, to advocate the exposure of human populations to radiation.

Many persons today might find it difficult to believe that the biological damage that can result from radiation exposure was once officially, and vehemently, denied. An article published in the March 25, 1955, issue of *U.S. News and World Report* concluded: "Scare stories now being circulated about grave dangers from current U.S. atomic tests are turning out to be without basis in fact." It also contained these paragraphs:

> *Statement*—Ray R. Lanier and Theodore Puck, scientists at University of Colorado: "For the first time in the history of the Nevada tests, the upsurge in radioactivity measured here within a matter of hours after the tests has become appreciable. It is not our desire to alarm the public mind needlessly, but we feel it is our duty to say so."
>
> *Answer*—Governor Edwin C. Johnson of Colorado called the Lanier-Puck report "phony" and said its authors "should be arrested." The governor added, "It will only alarm people. Someone has a screw loose some place, and I intend to find out about it."

Because the same article had claimed that exposure to radiation had given rise to a "race of fruit flies with increased vigor, hardiness, resistance to disease, and better reproductive ability," I became involved in a correspondence that was initiated by H. J. Muller and that eventually encompassed twenty or more geneticists. In a letter to Muller (April 5, 1955), having expressed my dismay at what was, in effect, a shallow analysis of the danger associated with radiation exposure, I added:

> I further believe, however, that we must not allow a fear of possible misinterpretation to inhibit research in population genetics—especially at the present time when such research is most urgently needed. As you know, some of the arguments concerning radiation damage have enormously wide ramifications in the fields of plant and animal breeding, speciation, and evolution. Research in any of these fields which tends to modify basic concepts upon which some aspects of radiation damage are predicted risks creating the erroneous impression that the conclusions regarding radiation damage are also wrong.

In his reply, Muller agreed that "we should not let the danger of such misrepresentations prevent us from continuing with valuable work along fundamental lines," adding, "and I would place your own work in that category." He added further: "Apparently there is a considerable area of disagreement between your and my conceptions of the processes of selection and evolution in man, and these cause our judgments concerning the long-range effect of radiation on human populations to be widely different, even though we are in general agreement concerning the radiation damage to individuals of generations not far removed from the exposed ones."

A final citation (B. P. Sonnenblick, remarks made at the forum "Atomic Energy in Industry," October 26, 1955, personal communication) illustrates how even competent scientists grossly oversimplified the issues involved and desperately sought convenient bases for resolving complex issues:

> You may have heard the names of Drs. Muller and Wallace, geneticists who disagree in terms of effects of radiation on populations. The former is a Nobel laureate; he is concerned, among other things, about the artificial restriction of the size of families of individuals in the upper socio-economic levels; their good genes are not passing into the population as frequently as they should. The question of what is a good gene is, of course, open to debate. Further, selection is relaxed. People do not die as readily or as quickly as in earlier periods and the genes which would have been eliminated persist and increase in the population gene pool. These add to the complexity of the general problem. Muller has been called an alarmist and a pessimist. I believe we would have to state that what might be now considered an alarmist view may some day be shown to have had a sound basis. Muller believes that under the conditions of modern society only a moderate increase in mutation rate can be tolerated, since civilized man may be near an equilibrium level whereby the elimination of many genes and the entrance into the population of other newly mutated genes are in balance.
>
> Muller's argument is that somewhere in a line of descent, any one of these genes will get into a particular genetic combination such that it will be eliminated because of the bearer's death or infertility. One or more mutant genes can tilt the balance adversely. Wallace, on the other hand, has reported recently on radiation experiments with fruit flies that after 120–130 generations, some lines are, if anything, superior in certain traits over those in untreated organisms. That appears good, but I think we had better not rush home yet and say, "Let us expose everyone to an excess of radiation. Everything will be fine."

4

RANDOM MUTATIONS AND VIABILITY

Data obtained from the study of irradiated *Drosophila* populations posed at least two difficult questions:

First, virtually from the moment of their discovery, X rays had been recognized as destructive agents. This was especially true for body tissues; many pioneer radiologists suffered radiation burns, developed cancerous growths, or died of leukemia. Muller's (1927) demonstration that X rays induced gene mutations was largely responsible for his later being awarded a Nobel Prize; the mutations he studied were recessive, sex-linked lethals. Despite the overwhelming evidence for the destructive effects of radiation-induced (or even spontaneous) mutations, the experimental populations had shown increases in fitness. How had they managed that?

Second, the rate at which the fitnesses of the chronically irradiated populations increased suggested that the responsible mutations were at least partially dominant; that is, they were retained in the population because of their (selectively favored) effects on heterozygous individuals. If this were the case, however, the populations must have suffered tremendous genetic (segregational) loads. How could such loads be tolerated?

Clearly, the focus of the experiments had to be shifted from the analysis of populations to a study of an array of randomly induced mutations with respect to their effects on their heterozygous carriers.

Fifty Years of Genetic Load

The deleterious effects of mutations on their homozygous (or hemizygous in the case of sex-linked mutations) carriers, in other words, was conceded; the *sorts* of chromosomes found in irradiated populations (not to mention decades of previous research) demanded that concession. Furthermore, the deleterious effect of the majority of lethal mutations on their heterozygous carriers was also conceded (see Simmons and Crow, 1977); this concession did not include *all* recessive lethals, however, because a small proportion seemed (perhaps, of course, because of linkage to nearby alleles) to improve the viability of their heterozygous carriers. Such variation among lethals was noted even in the first extensive study of the "recessiveness" of recessive lethals (Stern et al., 1952).

Had no previous study been designed to reveal the effect of an array of randomly induced mutations on their heterozygous carriers? With the exception of studies such as those dealing with lethal mutations or those with striking phenotypic effects, there were rather few. Those studies that had been carried out, if one conceded that heterozygosity among gene loci is common, may easily have compared the viability or longevity of, let's say, A_i/A_j with that of A_i/A^* or A_j/A^*, where A_i and A_j are naturally occurring alleles, and A^* is a radiation-induced mutation. To understand the effect of induced mutations, it is essential to compare A_i/A^* or A_j/A^* with either A_i/A_i or A_j/A_j individuals. The proper comparison requires carefully designed experiments; the mating of irradiated and nonirradiated control males with nonirradiated females for the purpose of comparing the two sets of progeny, for example, does not suffice.

The models

Two contrasting models are illustrated in Figures 4-1 and 4-2. The first (Figure 4-1) is based upon the prevailing belief of the 1950s (and earlier) that a wild-type, or "normal," allele occurs with high frequency at each gene locus; all other (and much rarer) alleles are mutations that lower the fitness of their carriers—especially that of homozygous carriers. Under this model, the ideal individual would carry only normal alleles at every locus (Figure 4-1a). Recurrent mutation, however, introduces mutant alleles into the population, where

(a) IDEAL GENOTYPE	A B C D E F G H ···								
	A B C D E F G H ···								
(b) "ORDINARY" GENOTYPE	A B C D e F G H ···								
	a B C D E F g H ···								
(c) ARTIFICIAL HOMOZYGOTE	A B C D e F G H ···								
	A B C D e F G H ···								
(d) ARTIFICIAL HOMOZYGOTE HETEROZYGOUS FOR NEWLY INDUCED MUTATION	A B C D e F G H ···								
	A B C D e F G h* ···								

Figure 4-1. Various aspects of the population model that postulates that there is but one normal allele at each gene locus (uppercase) and that all other alleles (lowercase) are to some extent deleterious. (a) The model states that the ideal individual would be entirely homozygous for uppercase letters. (b) The interaction of mutation and selection results in the presence of some deleterious alleles in a population; these appear as sporadic lowercase letters in the diagram. (c) The homozygous F_3 wild-type flies of the $Cy\ L$ and the $Cy\ L–Pm$ test cultures appear in the diagram as being homozygous even for rare excessive alleles (e). (d) Flies corresponding to those in part c except that during the mating procedure, one of the two chromosomes (*not* both) was irradiated (500-R X rays) with the result that the allele at one locus is shown as having mutated (allele H in c has been altered to h^* in d). (From Wallace, 1981, courtesy of Columbia University Press.)

they accumulate until their elimination by selection equals the rate at which they arise by mutation; as a consequence, an occasional gene locus is occupied by a mutant, rather than a normal, allele (Figure 4-1b).

The *ClB*-like procedures allow *Drosophila* geneticists to manipulate individual chromosomes virtually at will. The outcome of manipulating the second chromosomes of D. *melanogaster* by means of the $Cy\ L$ technique illustrated in Figure 3-1 is the production of homozygous wild-type individuals ($+_i/+_i$). Just what "homozygous" means in this case is shown in Figure 4-1c: every locus on the chromosome is, as a result of the special mating scheme, homozygous for the particular allele found at that locus. If, as shown in Figure 4-1c, the allele e were lethal when homozygous, the wild-type flies would not live. Because a chromosome possesses many gene loci, the probability that at least one locus will be occupied by a mutant allele (such as e in the figure) is

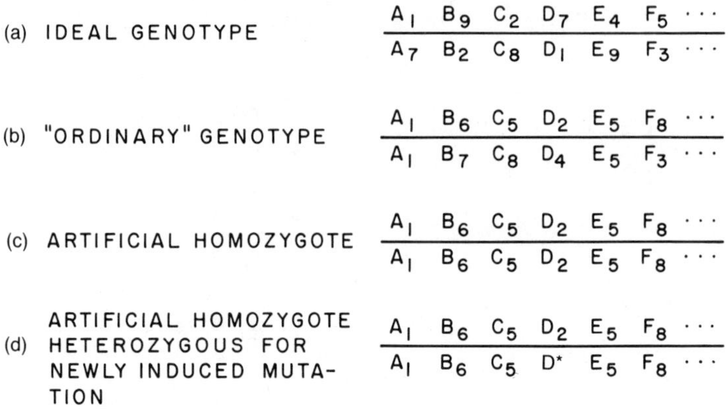

Fifty Years of Genetic Load

(a) IDEAL GENOTYPE	$\dfrac{A_1 \quad B_9 \quad C_2 \quad D_7 \quad E_4 \quad F_5}{A_7 \quad B_2 \quad C_8 \quad D_1 \quad E_9 \quad F_3}$	$\cdots$ $\cdots$
(b) "ORDINARY" GENOTYPE	$\dfrac{A_1 \quad B_6 \quad C_5 \quad D_2 \quad E_5 \quad F_8}{A_1 \quad B_7 \quad C_8 \quad D_4 \quad E_5 \quad F_3}$	$\cdots$ $\cdots$
(c) ARTIFICIAL HOMOZYGOTE	$\dfrac{A_1 \quad B_6 \quad C_5 \quad D_2 \quad E_5 \quad F_8}{A_1 \quad B_6 \quad C_5 \quad D_2 \quad E_5 \quad F_8}$	$\cdots$ $\cdots$
(d) ARTIFICIAL HOMOZYGOTE HETEROZYGOUS FOR NEWLY INDUCED MUTATION	$\dfrac{A_1 \quad B_6 \quad C_5 \quad D_2 \quad E_5 \quad F_8}{A_1 \quad B_6 \quad C_5 \quad D^* \quad E_5 \quad F_8}$	$\cdots$ $\cdots$

Figure 4-2. Various aspects of a population model postulating that the presence of two differing alleles at every locus is required for the optimal functioning of those genes. (a) The ideal individual would be heterozygous at all gene loci. (b) The chance loss of some alleles and the multiplication in frequency of others that would occur under Mendelian inheritance would lead to considerable homozygosis within the population (A_1/A_1 and E_5/E_5, for example). (c) The homozygous F_3 wild-type flies of the $Cy\ L$ and $Cy\ L$–Pm tests are homozygous at all loci on the tested chromosome. (d) Flies corresponding to those shown in part c except that during the mating procedure, one of the two chromosomes (*not* both) was irradiated (500-R X rays) with the result that heterozygosity has been partially restored (allele D_2 in c has been altered to D^* in d). (From Wallace, 1981, courtesy of Columbia University Press.)

high; few chromosomes, if any, would be expected to be mutant free. These mutant alleles, of course, are the bottlenecks that determine the viabilities of homozygous individuals.

Figure 4-1d shows that if one of the chromosomes (but not both) were exposed to radiation during the mating procedure, any radiation-induced mutation should impose additional harm on the already impaired viability of the homozygous flies. Recall that the majority of all gene loci are occupied by "normal" alleles. A properly designed experiment based on this model, then, would result in a comparison of H/H individuals with H/h^* ones, where h^* is a newly induced, harmful (by definition) mutant allele. Loci that remain unaffected despite the radiation exposure would be identical in the control and experimental flies, as has been shown in the case of loci C and D in the figure.

A second model, one postulating a higher fitness for heterozygous individuals, is illustrated in Figure 4-2. Here, heterozygosity at every locus has been represented (Figure 4-2a) as the ideal situation. This, of

Random Mutations and Viability

$$\text{CONTROL VIABILITY} \quad \frac{CyL}{Pm} : \frac{CyL}{+} : \frac{Pm}{+} : \frac{+}{+}$$
$$\qquad\qquad\qquad 1 \qquad 1-r \qquad 1-s \qquad 1-t$$

$$\text{EXPERIMENTAL VIABILITY} \quad \frac{CyL}{Pm} : \frac{CyL}{(+)} : \frac{Pm}{+} : \frac{+}{(+)}$$
$$\qquad\qquad\qquad 1 \qquad 1-r' \qquad 1-s' \qquad 1-t'$$

Figure 4-3. The four genotypes (see Figure 3-4) that reveal the effect of a single irradiated (500-R) chromosome on the viability of its otherwise homozygous carrier. In the $CyL-Pm$ test, the CyL/Pm flies serve as the standard (= 1.00) for comparison; flies of the other classes have relative viabilities that differ by amounts $r, r', s, s', t,$ or t'. The irradiated chromosome is shown in parentheses (+). Note that in comparing the viabilities of flies of corresponding genotypes in the two series, one assumes that the two standards (1.00 for the control and 1.00 for the experimental cultures) are, in fact, the same. (From Wallace, 1981, courtesy of Columbia University Press.)

course, represents the extreme case. Mendelian inheritance does not provide for the permanent retention of heterozygosity; hence, the ordinary individual, even under the extreme case illustrated here, would be homozygous at a considerable number of loci (Figure 4-2b).

The $Cy L$ procedures lead to F_3 test cultures in which the wild-type flies are homozygous for an entire chromosome $(+_i/+_i)$. Each locus in such flies would be homozygous, as shown in Figure 4-2c. As in the previous model, correspondingly homozygous flies could be produced, except that one chromosome (not both) was exposed to radiation. Again, the result is a comparison of a homozygote (D_2/D_2) with a newly created heterozygote (D_2/D^*). In this case, however, heterozygosity has in some sense and to some extent been restored. Unfortunately, the model does not lead to a prediction concerning the effect of this newly induced heterozygosity because one knows neither the proportion of alleles per locus that are advantageous when heterozygous nor the average magnitude of this advantage. Consequently, only the first model (Figure 4-1) can be used as a predictive model. According to this model, *all* mutations are harmful with respect to fitness. Furthermore, the majority of these mutations are expected to express their harmful effects even in heterozygous carriers. Radiation, therefore, should lower the egg-to-adult viability of the experimental flies relative to the control ones.

Figure 4-3 reveals that the $Cy L-Pm$ test procedure of Figure 3-4 was used in assessing the effect of irradiated (500 R) chromosomes

Fifty Years of Genetic Load

(and the presumed mutations they carried) on the viability of otherwise homozygous (second chromosome) wild-type flies. The complex mating scheme leading to these F_3 test cultures need not be given here (see Wallace, 1958 or 1959b). Figure 4-3 reveals that the relative viabilities of wild-type flies with (experimental) and without (control) irradiated chromosomes are determined relative to the $Cy\ L/Pm$ flies that appear in each test culture of both series of cultures; the wild-type flies of the two sets of cultures cannot be compared directly in these experiments.

The results of these experiments are presented in Tables 4-1, 4-2, and 4-3; they can be summarized as follows:

- Otherwise homozygous wild-type flies carrying an irradiated chromosome exhibited consistently higher average viabilities than their controls; because the sample sizes in the tests yielding these flies were large (more than 7600 cultures were examined in all), the difference, though small, is statistically significant (Table 4-1).
- As a test of the experimental technique, genetically marked flies (see Figure 4-3) that carried nonirradiated chromosomes in both the control and experimental series were compared; their viabilities were virtually identical, as expected (Table 4-2).

Table 4-1. The viability effects of newly induced mutations in heterozygous condition in *Drosophila melanogaster* otherwise homozygous for their second chromosomes. The irradiated wild-type chromosome is marked ('). The standard viability, 1.000, is that exhibited by $Cy\ L/Pm$ flies in these cultures. Total numbers of cultures tested are given in parentheses. (After Wallace, 1959b.)

	$+_1/+_1$		$+_1/+_1'$	
A	1.008	(766)	1.033	(764)
B	1.000	(676)	1.007	(672)
C	0.989	(636)	1.015	(637)
D	0.979	(596)	0.989	(598)
E	0.983	(639)	0.990	(637)
F	0.992	(499)	1.002	(496)
Total		3812		3804

Average difference: 1.5%
Error: 0.5%
Probability: 0.002

Random Mutations and Viability

Table 4-2. The viability effects of identical, nonirradiated wild-type chromosomes on genetically marked heterozygous carriers in the control and experimental F_3 test cultures. The standard viability, 1.000, is that exhibited by Cy L/Pm flies in these cultures. Total number of cultures tested are given in parentheses. (After Wallace, 1959b.)

	Pm/+		Pm/+	
A	1.146	(766)	1.137	(764)
B	1.139	(676)	1.140	(672)
C	1.137	(636)	1.145	(637)
D	1.143	(596)	1.136	(598)
G	1.215	(839)	1.222	(837)
	Cy L/+		Cy L/+	
E	1.127	(639)	1.125	(637)
F	1.189	(499)	1.201	(496)
AVERAGE				
unweighted	1.157	(4651)	1.158	(4641)
weighted	1.158	(4651)	1.159	(4641)

- Depending upon the details of mating procedures in the preliminary crosses, one or the other of the genetically marked flies (see Figure 4-3) carried an irradiated wild-type chromosome in the experimental series. Because they were not homozygous, one might expect that radiation-induced mutations would lower the viabilities of these flies. The results are ambivalent. Ostensibly, the Cy L/+ and Pm/+ flies carrying irradiated wild-type chromosomes exhibit higher viabilities than their counterparts in the control series. Statistically, however, this ostensible increase is not compelling (Table 4-3).

For those who are unaccustomed to experiments of the sort described here, some illustrative facts might be cited:

- Because the average number of flies per culture is somewhat greater than 300, 7600 cultures entailed nearly 2,500,000 flies that were classified, sorted, and counted.
- With four classes of flies (Figure 4-3) appearing in approximately equal numbers in each culture, a $1\frac{1}{2}$% increase in the frequency of wild-type flies (in tests where the average number of flies per culture is about 300) means that, on the average, each experimen-

Table 4-3. The viability effects of newly induced mutations on their genetically marked heterozygous carriers. The irradiated chromosome is marked ('). The standard viability, 1.000, is that exhibited by Cy L/Pm flies in these cultures. Total numbers of cultures tested are given in parentheses. (After Wallace, 1959b.)

	Cy L/+		Cy L/+'	
A	1.094	(766)	1.115	(764)
B	1.093	(676)	1.108	(672)
C	1.105	(636)	1.110	(637)
D	1.100	(596)	1.108	(698)
G	1.185	(839)	1.182	(837)
	Pm/+		Pm/+'	
E	1.127	(639)	1.125	(637)
F	1.189	(499)	1.201	(496)
AVERAGE				
unweighted	1.129	(4651)	1.136	(4641)
weighted	1.127	(4651)	1.135	(4641)

tal culture contained a *single* wild-type fly more than each control culture. This one fly, interestingly, can be nearly accounted for by Hiraizumi's (1965) observation that flies carrying an irradiated (500 R) chromosome (but that are otherwise homozygous) have an average development time (egg to adult) 1.2 hours shorter than their corresponding controls. Egg-to-adult viability as measured by either the *Cy L* or *Cy L–Pm* procedure includes rapid development as a component (enhancer) of "viability."

Despite the extreme care in coding and matching the control and experimental cultures in the experiments described above, the minute difference between the viabilities of the wild-type flies in these two series of cultures (experimental and control) might easily be ascribed to some unrecognized idiosyncrasy of the laboratory. Thus the confirmation of the early observations first by Terumi Mukai and Isao Yoshikawa (1964) and subsequently by Takeo Maruyama and James Crow (1975) was greatly appreciated. The latter used an elegant, statistically highly efficient technique that is described below.

The experimental procedure used by Maruyama and Crow is illus-

Random Mutations and Viability

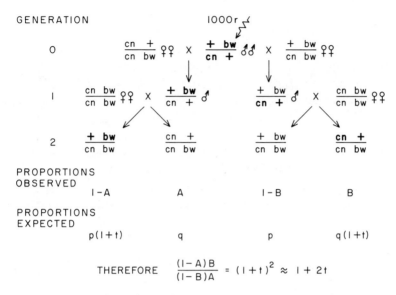

Figure 4-4. Mating scheme that leads in generation 2 to cultures in which the proportions of *cinnabar*- and *brown*-eyed flies reveal the effect of irradiated chromosomes (boldface) on the viability of their heterozygous carriers. An explanation of the algebraic exercises is given in the text. Note, however, that every fly counted in generation 2 is involved in the estimation of t, the effect of radiation exposure; this total involvement is the hallmark of an efficient experimental procedure. (After Maruyama and Crow, 1975, in Wallace, 1981, courtesy of Columbia University Press.)

trated in Figure 4-4. The genetic symbols are as follows: *cn*, cinnabar (bright red) eye color; *bw*, brown eye color; and +, the wild-type allele of each of these mutants. The alleles *cn* and *bw* are recessive; consequently, only homozygous individuals exhibit the abnormal eye colors. The double homozygotes (*cn bw*/*cn bw*) have white eyes (i.e., eyes that lack both brown and red pigments). This fact is useful for revealing the accidental presence of nonvirgin females in an occasional test culture; such cultures are then discarded.

As shown in Figure 4-4, phenotypically wild-type males carrying *cn* and *bw* on different chromosomes were exposed to 1000 R of X radiation. These irradiated males were mated, half with cinnabar-eyed females that carried *bw* in heterozygous condition and half with brown-eyed females that carried *cn* in heterozygous condition. In both series of crosses (left and right sides of Figure 4-4), phenotypically wild-type

Fifty Years of Genetic Load

males, each carrying an irradiated chromosome, were mated individually to *cn bw/cn bw* females.

The outcomes of the last cross in both series are cultures that yield cinnabar- and brown-eyed flies in proportions that might logically be designated p and q (rather than half and half), where $p + q = 1.00$. This expectation admits that the viabilities of these two types of flies need not be identical. However, the logical expectation is further modified by an amount t, in order to allow for an effect of an irradiated chromosome on its carrier's viability. How can this effect of radiation (t) be estimated? As shown in Figure 4-4, the observed proportions of *brown* and *cinnabar* flies are given as $1 - A$ and A in one series, and $1 - B$ and B in the other; these proportions are calculated from the experimental data themselves. The following equations can now be solved:

$$\frac{[(1-A)B]}{[(1-B)A]} = \frac{[p(1+t)q(1+t)]}{pq} = (1+t)^2.$$

If t is small, $(1 + t)^2$ equals approximately $1 + 2t$.

Maruyama and Crow's results are summarized in Table 4-4. The last line of the table deals with flies that correspond to those in Table 4-1, in which the effect of radiation-induced mutations in heterozygous condition on the viability of otherwise homozygous flies was found to be an increase of 1.5%. In Table 4-4 the effect was an increase of 1.6%. The first line of Table 4-4 deals with the effect of mutations in heterozygous condition on flies that (for unavoidable technical reasons) were possibly heterozygous at an unknown number of loci; here the newly induced mutations had an average deleterious effect, as expected. Finally, the center line shows that at an intermediate period between the early and late tests, the average effect of the radiation-induced mutations was nil—they neither increased nor decreased the relative viabilities of their carriers.

Explaining these data

The data cited above, when examined with respect to the two models that were presented earlier (Figures 4-1 and 4-2), suggest that the first

Table 4-4. The effect of newly induced mutations (1000 R) on the viability of their heterozygous carriers when these carriers are largely heterozygous, intermediate, or homozygous with respect to background genotype. (After Maruyama and Crow, 1975.)

Background	No. of flies	No. of cultures	Effect (t)
Heterozygous	540,913	4824	−0.017
Intermediate	460,908	5452	−0.000
Homozygous	456,956	4440	0.016

model should be rejected: a model that claims that all mutations are harmful in both homozygous and heterozygous conditions (barring complete recessivity) cannot readily account for an increased viability that accompanies an exposure of one of two chromosomes to radiation. Although the second model (Figure 4-2) was unable to predict the outcome of the experiments that were performed, the outcomes that were obtained by myself, Mukai and Yoshikawa, and Maruyama and Crow were compatible with that model.

After seeing the early results, one colleague suggested that perhaps the experimental results were being misinterpreted; perhaps, he suggested, low numbers of wild-type flies in F_3 test cultures indicated high fitness and high numbers reflected low fitness. This suggestion was abandoned when I asked whether he would have supported that interpretation had the results been otherwise. And what about the relative standings of the experimental populations? In each case, his response was no.

Dominance, however, represents a different matter. A second colleague suggested that "beneficial" mutations may be dominant in their effects on viability, while mutations that lower the viabilities of their homozygous carriers may be recessive. Precisely this argument has been used recently by Charlotte Paquin and Julian Adams (1983) in a study of genetic changeover in asexual haploid and diploid yeast populations. The diploid, they argue, might undergo more rapid changeovers because selectively favorable mutations can arise on either of the two genomes. Actually, the favorable mutations seemingly arose in their diploid yeast populations 1.6 times as frequently as in haploid ones, a finding that suggests that only 20% of all favorable mutations are recessive.

Fifty Years of Genetic Load

One reply to the suggestion that the results obtained by studying the viability effects of irradiated chromosomes might be explained by the dominance of viability-enhanced mutations (and the recessiveness of deleterious ones) was based on calculations that were standard for the times:

Let u equal the mutation rate per locus to recessive alleles whose effect is to lower the fitness of their homozygous carriers by an amount s. The frequency of such alleles at equilibrium will equal $\hat{q}$, or $\sqrt{u/s}$. This also equals the proportion of loci on a chromosome at which such alleles are found. In order to mutate one of these alleles (on the average) back to its normal state ($a \to A$), an average of $1/\sqrt{u/s}$ loci must be mutated. These mutations would increase the viability of the otherwise homozygotes by the amount s. If, in fact, it is known that individuals carrying the irradiated chromosomes have had their fitnesses increased by an amount d, then $d/s \times 1/\sqrt{u/s}$ mutations must have occurred. The number of mutations required to produce an effect d thus equals $d/\sqrt{us}$.

The above calculation is based on at least two unrealistic assumptions whose effect is to minimize the number of mutations required. First, every mutation of a deleterious allele has been said to restore the normal allele; the possibility that mutations to equally deleterious (or worse) alleles may occur has been ignored. Second, all mutations of normal alleles to deleterious ones must be assumed to be *absolutely* recessive; any degree of dominance of these newly induced mutations would require even more mutations to produce the enhancing effect of irradiated chromosomes. All that aside, however, letting $d = 0.015$, $s = 0.01$, and $u = 10^{-6}$, we find that an average of 150 mutations per chromosome are required to account for the experimental observations listed in Table 4-1. Larger numbers would be required for smaller values of s. The numbers of required mutations, considering the low level of radiation exposure, appear to be unrealistic.

A second reply, one that superficially seems to accept the notion that beneficial changes are dominant whereas deleterious ones tend to be recessive, stems from speculations on the structure of gene control regions (see Figure 4-5). The bit of chromosomal DNA now known to correspond to the Mendelian gene (*Mendelizing unit* may be a better term) contains regions that carry out a variety of functions; for example, the structural gene specifies the amino acid sequence of a poly-

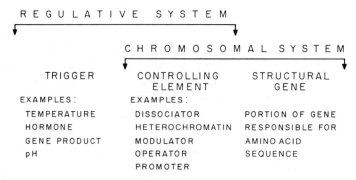

Figure 4-5. The dual nature of gene control systems (Wallace, 1963c): the system in its entirety provides for the "internalization" of the environment. One portion is entirely chromosomal; it consists of the structural gene and its associated enhancers, promoters, operators, transcription initiation sites, introns, exons, and other portions of nearby DNA. This assembly is essentially *the* Mendelian gene, although the region can be dissected for study by mutagenesis and high-resolution recombination analyses. The second portion, which terminates in the controlling region of the Mendelian gene, may originate in the external environment: hours of daylight, temperature, humidity, or organic compounds (pheromones) that reveal the presence of other organisms. R. J. MacIntyre (1982) has suggested that this is one of the first accounts that emphasized the control of gene action as a source of intra- and interpopulational heterosis. The five McClintock publications cited in this paper suggest that full credit could lie elsewhere. (From Wallace, 1981, courtesy of Columbia University Press.)

peptide, the transcription initiation sites pinpoint the start of transcription, and enhancers modulate the rate at which the structural gene is transcribed. The last two mentioned are but two of numerous *cis*-acting controlling elements, some of which *must* be located upstream of the structural gene as shown in Figure 4-5; others, however, can be downstream or even within the area occupied by the structural gene because the mRNA transcribed from the latter is often cut and spliced before translation, thus leaving regions (introns) within the gene's DNA untranslated.

R. J. Britten and E. H. Davidson (1969) postulated in one of two models that the upstream portion of the Mendelizing unit contains numerous receptor sites, each capable of receiving a signal from one of many coordinating control loci; the latter, because of corresponding receptors at many gene loci, may activate these diverse loci virtually simultaneously. Francis Crick (1971) extended the Britten-Davidson model by suggesting how sensor regions might be made to utilize single-stranded, rather than double-stranded, DNA as efficient signal-reading sensors.

Continuing this line of reasoning, T. L. Kass and I suggested further that the sequence of sensors preceding each structural gene might operate more efficiently if the precise linear orders of sensors for two homologous alleles were not identical but, instead, were even randomly shuffled relative to one another (Wallace and Kass, 1974). The rationale for this suggestion was that the rapid response of either allele to a particular call for action would best satisfy that call; in response to some calls, one allele would be the first to respond, whereas in response to others the homologous allele would respond first. This pattern of nonsimultaneous (asynchronous) response, in fact, could explain why random mutations in heterozygous condition might increase the average viability of their otherwise homozygous carriers (Figure 4-6). It is not a specific order in which the sensors at a gene locus are arranged that confers highest viability but, rather, the differences in the linear orders of sensors associated with pairs of homologous alleles.

I. M. Lerner (1954), with no clear model of gene structure in mind, suggested that homozygosis per se might be harmful to normally outcrossing diploid organisms. Both Muller (1959) and Kimura (1983: 17) have referred to this view as "pre-Mendelian." A reexamination of Lerner's argument in the light of molecular genetics might reveal that the "bead-on-a-string" view of the gene, the prevalent view of the 1950s and 1960s, was misleading; there is no bead—the gene is an elongated region of the string itself, whose different portions serve different functions. Lerner, in retrospect, may have been more post-Mullerian than pre-Mendelian.

The model Kass and I proposed (1974; also see Wallace, 1976), as stated earlier, was a derivative of Britten and Davidson's (1969) and Crick's (1971) models. Twice recently, both times under cover of a peer reviewer's anonymity, I have read that the Britten-Davidson model is obsolete; inquiries posed of molecular geneticists in personal conversations have not been answered nearly so bluntly. A brief review of matters as they seem to stand today may be of some value.

Citations of the Britten-Davidson model in the literature have by no means ceased. From the outset, however, a few molecular geneticists were not impressed by the Britten-Davidson model(s) for logical rea-

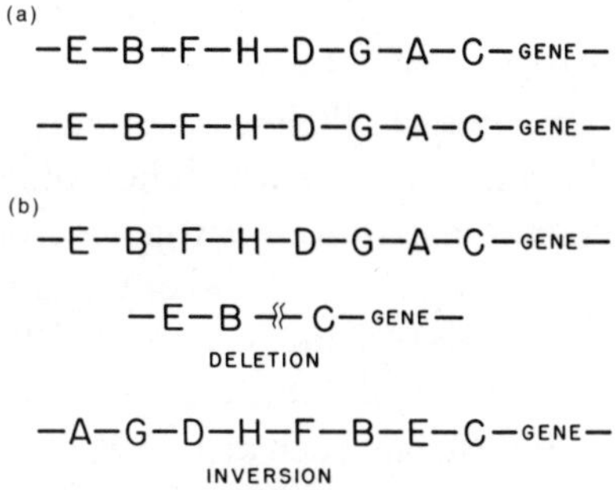

Figure 4-6. An attempt to explain the average *enhancing* effect of random mutations tested in heterozygous condition but in otherwise (largely) homozygous individuals, based on the Britten-Davidson model of gene control. The function of the structural gene (identified as "gene") is assumed to be controlled by many upstream sensors (E, B, F, etc.), each capable of receiving a signal from an appropriate regulator gene whose signal coordinates the action of this locus and several (or many) others. In homozygous individuals (a), the sequences of sensors are identical. Following irradiation, the sequence in the irradiated chromosome may be altered (b) by a small deletion or inversion. If the altered sequence improves the joint functioning of the two alleles at this gene locus, viability is enhanced; if not, the nonirradiated sequence continues to perform the functions of this gene locus. (From Wallace, 1981, courtesy of Columbia University Press.)

(a)
—E—B—F—H—D—G—A—C—GENE—

—E—B—F—H—D—G—A—C—GENE—

(b)
—E—B—F—H—D—G—A—C—GENE—

—E—B—⫫—C—GENE—
DELETION

—A—G—D—H—F—B—E—C—GENE—
INVERSION

sons alone. The model, they reasoned, propounded the obvious: one regulator can control many genes; any gene is under the control of many regulators. Anyone denying this general statement, the reasoning goes, must believe instead that each gene is controlled by one regulator, and that that single regulator is controlled by a more remote regulator, and so on without end. At some point this alternative system must collapse and allow one gene to have more than one regulator, and one regulator to control more than one gene.

The Wallace-Kass model merely emphasized that the two homologous alleles at a locus, by having dissimilar sequential arrangements of signal receptors (sensors) upstream of the homologous structural genes, would operate at slightly different times when called upon to act. If either of the two times was more appropriate than the other, the allele with the appropriate timing (rapid response to an incoming signal, perhaps) would be the effective allele for that moment. Because the dissimilarity of upstream control regions was said to arise by intrachromosomal rearrangements of DNA, the signal receptors were presumed to be embedded within a series of overlapping reverse repeats or extensive palindromes; recombination between reverse-oriented segments of DNA inverts the intervening region while leaving the end regions unaltered. Asynchrony of gene action was achieved in the Wallace-Kass model by mechanical means. The proposed structure for the DNA of control regions, however, has not been verified.

The original point of the Wallace-Kass model—asynchronous gene action—should not be summarily dismissed: the efficient action of homologous alleles, according to the model, lies in the somewhat different times at which homologous alleles react to signals. The successful insertion of *lux* genes from *Vibrio fischeri* into *E. coli* (Engebrecht et al., 1985) and of the luciferase gene from a firefly into cells of the tobacco plant (Ow et al., 1986) has inaugurated an entirely new approach to the study of the timing of gene action: a light reveals when luciferase has been synthesized—that is, when the gene has been activated (see Schauer et al., 1988). K. V. Wood et al. (1989) have succeeded in transferring a number of luciferase genes from the click beetle *Pyrophorus plagiophthalamus* into *E. coli*. These genes code for luciferases of four types distinguishable by the colors they generate: green (546 nm), yellow-green (560 nm), yellow (578 nm), and orange (593 nm). As these authors state, "Because of the different colors, these clones may be useful in experiments in which multiple reporter genes [read "alleles"] are needed." A definitive test of the Wallace-Kass model is virtually at hand!

Still another explanation for the improvement of the viability of homozygotes through the exposure of one chromosome to a low dose of radiation postulates a partial return to an *optimal level* of heterozygosity within the individuals carrying the irradiated chromosome. Why there should be an optimal level and what constitutes an optimal level are matters this argument does not make clear. My response to this (early) suggestion was as follows: The radiation exposure that I employed was not aimed at particular loci; thus, if there is an optimal level of heterozygosity, heterozygosity at any locus seemingly contributes to it. In that case, however, there is no reason to believe that geographically isolated populations of *Drosophila melanogaster* would utilize identical loci in meeting the optimal level within local populations. If the optimum level is a small percentage of all loci, interpopulation F_1 hybrids might easily have a level of heterozygosity nearly twice the optimum; consequently, such hybrids should often exhibit "negative" heterosis. They do not do so; F_1 interpopulation hybrids nearly always exhibit heterosis with respect to many fitness traits (see Wallace and Vetukhiv, 1955).

Recently, considerable emphasis has been placed on the dysgenesis observed among the progeny of interpopulation hybrid *Drosophila melanogaster* (Kidwell et al., 1977; Woodruff and Thompson, 1980). My impression, however, is that the observed level of dysgenesis is low compared with the commonly observed increase in average fitness (measured as the egg-to-adult or first-instar-larva-to-adult survival either in pure culture or in the presence of a genetically marked competitor) of interpopulation hybrids. In fact, recalling my early experiments, it seems to me that the experimental technique actually penalized these hybrids. The dessication of an occasional vial (these were much smaller than the standard 25 mm × 95 mm shell vial) would often cause the deaths of all F_1 interpopulation hybrid larvae (because of their uniform development), whereas one or a few adults would emerge from dessicated vials in which the sizes of progeny larvae were not uniform. Visual inspection of individual cultures before they became dessicated always revealed vigorous larvae of uniform size in the cultures of interpopulation F_1 hybrids, but larvae of variable sizes (even small dead larvae) in cultures where recombination had occurred in the mother or in the mothers of the parental males. (Recombination does not occur in male *D. melanogaster*.)

This account can be concluded by saying that not all experiments of the sort we have been discussing (namely, the effect on viability of mutations in heterozygous condition but in otherwise homozygous background) have led to positive results. The results that have been statistically significant, however, have been positive. The literature for this research has been reviewed rather extensively (Wallace, 1981: chapter 14). One extensive set of data that has not yet been reviewed adequately, however, is that collected by Mukai and his colleagues; this important research spans more than twenty years and appears in more than a dozen research papers (see Mukai, 1985). These studies are so extensive that they deserve a separate, comprehensive analysis. Two caveats may be appropriate, however, for anyone who plans to undertake such a review and synthesis:

- In order to study as many cultures as possible, Mukai and his colleagues have frequently resorted to the "two-class" *Cy L* technique; the frequency of wild-type flies in such cultures is affected by dominance, especially in the case of homozygous tests where observed frequency and actual fitness may be inversely related.
- In my experience, individual wild-type chromosomes may interact

idiosyncratically with genetically marked ones. For reasons other than sheer numbers, averages obtained from analyses of many chromosomes are more reliable, as one might expect—such analyses "average over" gene interactions. This caveat applies especially to attempts at identifying particular individual cultures as containing either "normal" or "mutant" chromosomes. Apparent decreases in the viability of wild-type homozygotes that occur over time, for example, may reflect either an accumulation of deleterious mutations in the wild-type chromosome or selection for improved viability of $Cy\ L/+$ heterozygotes (Wallace, 1965).

Personal comments

The manuscript dealing with the viability effects of random mutations that eventually appeared in *Evolution* (Wallace, 1958) was submitted first to *Genetics* even though it obviously exceeded the lengths of papers allowed by that journal. I asked the editor to consult the editorial board about page charges; if there were to be charges, I asked that the manuscript be returned without review because I had no funds for publication costs. Things went otherwise. The manuscript was sent out for review before the matter of page charges was looked into. Upon hearing that there would be a charge, I requested once more that the manuscript be returned.

Along with the returned manuscript was the first of two reviews received by the editor of *Genetics*. Its gist was that hybrid vigor is an ancient topic, that my results were to be expected, and that I had contributed nothing new, except perhaps to demonstrate that special *Drosophila* techniques could be applied to the study of hybrid vigor. So much for the two models described in Figures 4-1 and 4-2! My depression ended several days later when the second review (signed by Sewall Wright) arrived. Its tone was entirely different from that of the first; in essence, Wright said that the results, if they could be confirmed, were important ones.

As was the custom of many authors in the 1950s, I had sent preliminary copies of the manuscript to several colleagues, including H. J. Muller, in order to obtain criticisms and suggestions before submitting it for publication. Years later, one of Muller's ex-students related this account of the arrival of my paper at Muller's laboratory:

At one of the group's lunchtime journal club meetings, Muller pre-empted the program in order to review a manuscript he had just received. At the end of his presentation, he confronted the group with a challenge, "Tell me, what's wrong with this paper?"

The students, as Muller's students were wont to do, looked to Irwin Herskowitz, a young staff member, to respond for them. After several false starts, Herskowitz said, "I don't see anything wrong with it."

"Neither do I," Muller reportedly replied.

5

DILEMMAS AND OPTIONS

The study of the viability effects of newly induced mutations in heterozygous condition but in an otherwise homozygous background had been undertaken to answer a question that the experimental *Drosophila* populations had raised but were incapable of answering. Mutations advantageous to their heterozygous carriers had apparently arisen in the irradiated populations; as a result, the fitnesses of the experimental populations increased relative to that of the control population even as the frequencies of lethals, semilethals, and chromosomes with detrimental effects on viability increased in those populations (numbers 5, 6, and 7) that were irradiated continuously. One could then ask, What proportion of all new mutations have this selectively favorable effect on their heterozygous carriers?

That the average viability of flies carrying new mutations in heterozygous condition exceeded that of their otherwise homozygous controls seemed to provide an answer: chromosomes, when made homozygous (at every locus, remember) by special crossing procedures, seem to result in viabilities whose average is *in*creased when one of the two identical alleles at one or another of many loci is mutated to a different state. This empirical observation suggested to me that, at least to an extent compatible with population size and the chance loss of alleles at one or another locus within the population, more than half of all loci would exhibit allelic polymorphisms—perhaps considerably more than half. (In retrospect, it appears that this conclusion was based on the assumption that the advantage conferred by certain newly induced heterozygosities was approximately equal to the disadvantages that may have been conferred by others.)

The matter does not terminate with that inference, however. By what mechanism does a randomly induced mutation result in an improvement in viability—an improvement in what is already the organism's ability to undergo seemingly normal development from early embryonic stages through pupation and eclosion? It may be recalled that the homozygous control flies of my experiments were themselves quasi-normal in viability. R. A. Fisher, who visited the laboratory at Cold Spring Harbor at that time, was especially unhelpful: "A watch that runs too slowly is often improved by a gentle shaking," was his only comment. Even Maruyama and Crow (1975), seventeen years after my original observations, noted that while the results of three workers agreed, the physiological mechanism by which the improvement occurred was not at all clear. In 1958–1959 a mechanism explaining the observed viability effects appeared as remote as one explaining perpetual motion. Consequently, a continuation of this line of research on the part of a population geneticist seemed futile.

What, then, were the logical directions that future research might take? The study of allozyme variation in populations was one excellent possibility. Even before his paper appeared in press (Wright, 1963), T. F. R. Wright, a young colleague at Cold Spring Harbor, had shown me starch gels with stained bands revealing the slow, fast, and both slow and fast bands of flies homozygous and heterozygous for the "slow" and "fast" alleles at the Esterase-6 locus. He had also shown me data (Table 5-1) demonstrating that these enzyme molecules exhibited Mendelian inheritance. Wright showed little interest in the possibility that a survey of many different enzymes might reveal a large proportion with similar polymorphisms; instead, he proposed to concentrate on the properties of the two forms of Esterase-6 in an effort to learn the basis for this particular polymorphism's existence. Because I lacked any training in or a particular love for biochemistry and its exacting laboratory routines (training and a love that I believe are essential for one undertaking a study of enzyme variation), I decided to look elsewhere for research problems.

There remained another seeming contradiction stemming from the

Table 5-1. An analysis of the inheritance of fast and slow Esterase-6 bands in *Drosophila melanogaster*. Parental flies were squashed and subjected to electrophoretic analysis after they had produced progeny larvae; the latter were analyzed after they had become adults. (After T. R. F. Wright, 1963.)

	Offspring		
Parents	FF	FS	SS
FF × FF		not tested	
FF × FS	20	21	0
FF × SS	0	59	0
FS × FS	14	25	7
FS × SS	0	39	43
SS × SS	0	0	12

studies of irradiated *Drosophila* populations, and especially from the study of the viability effects of randomly induced mutations on their heterozygous carriers: if the selective advantage of heterozygotes was as common as the data appeared to suggest, the resulting genetic (segregational) load should be too large for the population to bear. To postulate (Wallace, 1963b) the existence of many alleles per locus, while providing an escape from the load problem, was not satisfactory. Where, for example, was the evidence for these many alleles? Hemoglobin was the best-known protein of that era, and, as one worker had expressed it, all normal persons have normal hemoglobin. What, then, is the relationship between genetic load and the existence or extinction of a population?

Kimura and Crow (1964) published an important paper on this relationship at just that time. Haldane had published a paper in 1957 on the cost of natural selection. Because these two papers exerted a strong influence on Kimura (1968) and his formulation of the neutral theory, both are reviewed in the following pages; the paper by Kimura and Crow (1964), however, relates more closely to the data discussed in the preceding chapters.

Kimura and Crow (1964)

In this paper Kimura and Crow solve two problems that bear closely on the matter of the irradiated populations and the subsequent, more "focused," experiments. One problem deals with the possible ubiquitous advantage of heterozygous individuals; the other with the level of genic variation that might be expected to exist in populations.

Let's assume that at any gene locus (for example, the *a* locus) there exists an essentially infinite number of alleles: $a_1, a_2, a_3, \ldots, a_i, a_j, \ldots, a_\infty$. Every mutation that occurs at this locus creates an allele that has never existed before. Furthermore, let's assume that heterozygotes for any pair of alleles (a_i/a_j) have an average fitness exceeding that of the corresponding homozygotes (a_i/a_i and a_j/a_j). Finally, let the fitness of all heterozygous individuals be equal (and equal to 1.00) and those of all homozygotes also be equal (and equal to 0.99).

The calculations that follow from these assumptions are complex and can be consulted in the original paper. Here, I shall only summarize the consequences of the numerical values alluded to parenthetically above.

- The greater the number of alleles existing at a locus, the smaller the frequency of each in the population. Furthermore, the smaller the frequency of alleles, the more easily they are lost by chance from the population. With the fitnesses that have been specified above, and with the enormous number of possible alleles, a balance between mutation to new alleles ($u = 10^{-5}$) and the loss of old ones in a population of 10,000 individuals occurs when the effective number of alleles per locus is eight.
- The frequency (F) of homozygotes at a locus is (at least approximately) the reciprocal of the effective number of alleles; thus, with eight alleles, one-eighth of the individuals are expected to be homozygous for one or the other of these alleles. The genetic load created by this locus, then, equals $\frac{1}{8} \times 0.01$ ($= Fs$) $= 0.00125$.
- If as many as 5000 loci possessed a system of independently acting, mutually heterotic alleles such as that described here (recall that I had suggested that more than one-half of all gene loci might be occupied by alleles whose effect on viability is improved by being heterozygous with a dissimilar allele), the average fitness of the population would be $(0.99875)^{5000}$, or 0.002, relative to that of the optimum genotype (1.00). The calculated genetic (i.e., segregational) load, 0.998, appears to be larger than a population—even a population of flies where females lay hundreds of eggs—might be expected to bear.

Fifty Years of Genetic Load

The other model discussed by Kimura and Crow also postulates an enormous number of possible alleles—so large a number that each new mutations is to an allele not preexisting in the population. The important phenomenon in this model is *inbreeding,* the bringing together of alleles that are identical by descent from one that existed in an earlier generation. In a population of N individuals (N is really the *effective* number of individuals, a number that is generally somewhat smaller than the number of adults that might be counted during a census), the probability that two gametes carrying alleles identical by descent from the previous generation equals $\frac{1}{2N}$; the probability that uniting gametes will *not* carry alleles that are identical by descent equals $\left(1 - \frac{1}{2N}\right)$, or $\frac{(2N-1)}{2N}$.

The probability that two uniting gametes of generation t will carry alleles descended from a common ancestor of t generations ago equals

$$F_t = \frac{1}{2N} + \left[\frac{(2N-1)}{2N}\right]F_{t-1}.$$

The two alleles will be identical in physical state as well as by descent only if neither has mutated during the previous generation. In this sense,

$$F_t = \left\{\frac{1}{2N} + \left[\frac{(2N-1)}{2N}\right]F_{t-1}\right\}(1-u)^2.$$

At equilibrium, $F_{t-1} = F_t = F$. Furthermore, $(1 - u)^2$ is approximately equal to $(1 - 2u)$. Consequently, the equation reduces to (approximately)

$$F = \frac{1}{(4Nu + 1)}.$$

The effective number of alleles equals $1/F$, or $4Nu + 1$.

The effective number of alleles expected per locus under different mutation rates and population sizes are summarized in Table 5-2. If population size is large (10^5 or 10^7) and mutation rate is high (10^{-4} or

Dilemmas and Options

Table 5-2. The average proportion of homozygosity (upper figure) and the effective number of alleles per locus (lower figure) in a randomly mating population of effective size N_e. The alleles are selectively neutral and the mutation rate of any allele is u. The number of possible mutant states is assumed to be large enough so that each new mutant is different from the others in the population. (Kimura and Crow, 1964.)

Mutation rate, u	Effective population number, N_e					
	10^2	10^3	10^4	10^5	10^6	10^7
10^{-4}	.96	.71	.20	.024	.0025	.00025
	1.04	1.4	5.0	41	401	4001
10^{-5}	.996	.96	.71	.20	.024	.0025
	1.004	1.04	1.4	5.0	41	401
10^{-6}	.9996	.996	.96	.71	.20	.024
	1.0004	1.004	1.04	1.4	5.0	41
10^{-7}	.99996	.9996	.996	.96	.71	.20
	1.00004	1.0004	1.004	1.04	1.4	5.0

10^{-5}), the number of alleles that might be found at a locus can be very large indeed. However, if populations tend to number 1000 to 10,000 (as Kimura and Crow tended to believe at the time), then, under mutation rates of 10^{-5} or 10^{-6}, the effective number of alleles is very nearly 1.00.

The two analyses outlined in the preceding paragraphs led to the following conclusions: (1) a selective superiority of heterozygotes, if it were to occur at many gene loci, would lead to an unbearable genetic load; and (2) the loss of alleles by chance alone under seemingly reasonable values of u and N would reduce the effective number of alleles per locus virtually to 1.00. Consequently, populations probably do not harbor extensive polymorphisms of any sort; on the contrary, their members are essentially homogeneous in genetic composition. There was a caveat in the 1964 paper, however, that was based on the upper right corner of Table 5-2: if population sizes are large and mutation to nearly neutral alleles is frequent, many alleles could exist in a population. Later, when previously undetected (and unsuspected) variation was revealed by newly developed molecular techniques (for example, Lewontin and Hubby, 1966), the earlier view concerning effective population size quickly changed.

The cost of natural selection: Haldane (1957)

Haldane's calculations on the cost of natural selection are discussed at this time not because they bear directly on the question of genic polymorphisms in populations but because they constitute one of the bases upon which the asserted need for the neutral theory rests (Kimura, 1968). After presenting Haldane's calculations and conclusions, this chapter will conclude with some comments of my own concerning (as I see it) the danger that accompanies the adoption of and subsequent strict adherence to mathematical conventions that were originally invoked for computational convenience. The following account of Haldane's calculations is only slightly changed from one I gave earlier (Wallace, 1981:304ff.).

The replacement of one allele by another in a population, unless it happens solely as the result of chance events (as for neutral alleles), requires that the carriers of the two alleles differ in fitness—that is, in either survival, reproductive success, developmental rate, or some other component of Darwinian fitness. For simplicity, we will follow the account given by Haldane (1957) for a haploid organism. Calculations for diploid organisms are considerably more complex but nevertheless lead to similar conclusions.

In the following outline, the relative fitnesses of A and a (haploid) individuals are equal to $1 - k$ and 1.00, respectively. The inferior fitness of A individuals is the outcome of a recent environmental change; consequently, although they possess the lower fitness, A individuals are the more common: their frequency, q, is nearly equal to 1.00. Inversely, the frequency of a individuals, p, is virtually zero. The consequences of these stipulations are as follows:

1. The fraction of selective deaths in any generation, n, equals kq_n.
2. The frequency of A in any generation, $n + 1$, equals

$$q_{n+1} = \frac{q_n - kq_n}{1 - kq_n}.$$

3. The change in gene frequency, $\Delta q_n = q_{n+1} - q_n$, equals

$$-\frac{kq_n(1 - q_n)}{1 - kq_n}.$$

4. The total fraction of selective deaths equals

$$D = k \sum_{n=0}^{\infty} q_n.$$

5. When k is small,

$$\frac{dq}{dt} = -kq(1-q).$$

6. By integration,

$$D = k \int_0^{\infty} q \, dt,$$

or

$$D = k \int_0^{q_0} -q \frac{dt}{dq} dq,$$

but

$$\frac{dt}{dq} = -\frac{1}{kq(1-q)};$$

therefore

$$D = \int_0^{q_0} \frac{dq}{1-q} = -\ln(1-q_0) = -\ln p_0.$$

The cost of natural selection (that is, the total fraction of selective deaths occurring as a result of the displacement of one allele [the common one] by another [the rare one]) is independent of the intensity of selection; instead, it depends entirely upon the initial frequency of the rare, but now advantageous, allele. Initial frequencies (p_0) of 0.001, 0.0001, and 0.00001 lead to total selective deaths 6.91, 9.21, and 11.51 times the normal population size, respectively. Haldane suggested that perhaps only 10% of a population's reproductive effort could be expended on the replacement of one allele by another; as a consequence, this type of evolutionary change must be a slow process. Using the above figures for selective deaths, the replacement of A by a would require from 69 to 115 generations. For diploid organisms, Haldane suggested 300 generations as a reasonable length of time required for the substitution of one allele by another.

I believe that the above calculations are flawed. If there is only a *single* individual of the rare, advantageous type, that type can be assigned lower and lower frequencies only by assuming that it is a member of larger and larger populations; for example, $p_0 = 0.001$ suggests that the population consists of 1000 individuals, whereas $p_0 = 0.00001$ suggests a total population of 100,000 individuals. For a given k (Haldane's symbol for selection coefficient), more generations will be required for the favored allele to replace its (now-disfavored) counterpart in a large population than in a small population. Haldane (1924) himself showed that the number of generations required for a given change in gene frequency is inversely proportional to the intensity of selection (see Wallace, 1981:310, for a corresponding trigonometric inference); hence, the product of the selection coefficient times the number of generations required for the displacement of one allele by another is a constant. That constant is essentially what Haldane calculated as "selective deaths."

The arbitrary convention that states that the maximum, or optimal, fitness shall be assigned the value 1.00 is also responsible for the conclusions (erroneous, in my view) reached by Haldane concerning the cost of natural selection. This claim is best illustrated by a numerical example. Imagine a population consisting of 100,000,001 asexual organisms of which 1 (A) has a 1% selective advantage over the others (a). By convention, the fitness of A is set at 1.00 and that of the a individuals at 0.99 ($= 1 - k$). After selection has operated, 1,000,000 a individuals are said to have died. Normalization of the population to its original size, however, results in the resurrection of most of these "dead" individuals: after normalization, the population contains 1.01 A individuals and 99,999,999.99 a ones. In the next generation, virtually 1,000,000 a individuals are said to die again. Haldane's selective deaths represent the sum of these calculated deaths; the corresponding resurrections are ignored.

Dilemmas and Options

Consider the result obtained by setting the relative fitnesses of *A* and *a* at 1.01 and 1.00, respectively. Here, it is best to imagine 100 identical populations in 100 culture tubes each consisting originally of 100,000,000 *a* individuals and a single *A*. After selection, 99 of the 100 populations will have remained precisely as they were at the start; in the remaining population there were be 2 *A* individuals. In all, there are now 101 *A* individuals rather than the 100 there were at the start. That is what fitness 1.01 means! However, if the carrying capacity of the environment within each culture tube is so precise the none can support 100,000,002 individuals, then an *a* individual must be sacrificed in the tube containing two *A* individuals. The increase in the average number of *A* from 1 to 1.01 per population has been accomplished by the elimination (that is, at a cost) of a single *a* individual in 100 populations, or an average of 0.01 *a* individuals per population.

The replacement of *a* individuals by *A* individuals under the alternative scheme presented here will call for the elimination of one *a* individual for each additional *A*; the total number of *a* individuals eventually eliminated will equal the total number that originally existed: no more, no less. The seemingly exorbitant cost of natural selection calculated by Haldane can be ascribed either to his (unnecessary) adherence to a computational convention (note how clearly the second example reveals events as they would actually occur) or to a misuse of an accepted convention (his failure to realize that normalization of the population to 100% while maintaining a constant population size resurrected most "dead" individuals from the category designated "selective deaths").

Personal comments

William Feller, a mathematician at Princeton University and author of *An Introduction to Probability Theory and Its Applications* (1950), was made aware of Haldane's paper, "The Cost of Natural Selection," by his friend Th. Dobzhansky. Having examined the problem, Feller (1967) concluded that Haldane had erred and that the number of selective deaths occurring during the substitution of one allele by another equals the number of individuals that originally carried the now-disfavored allele. This is the conclusion that has been reached above.

Kimura and Crow (1969) and Kimura and Ohta (1971) have criticized Feller's conclusion. Kimura and Ohta (1971:74) state the basis for the criticism rather clearly: "Feller overlooked an important biological fact that in the process of gene substitution in evolution, the relative proportions of genes change enormously, *while the total population number remains relatively constant* due to population regulating mechanisms." Later they say: "It is often not realized how important is the power of natural selection by which a more advantageous gene can gradually supplant less advantageous genes in a population *without appreciably affecting the total population number*. By this power, an advantageous gene combination that was initially very rare or even non-existent can finally emerge as the prevailing type in a population whose total number is always restricted by the carrying capacity of the environment." (Emphasis mine in both instances.)

In addition to their enormous contributions to mathematical genetics, both Kimura and Crow are biologists by training. Kimura's career, according to the late Ken-ichi Kojima, began with a study of handedness in wheat; Crow once studied the viability of *Drosophila* hybrids and ether sensitivity in several *Drosophila* species. I mention these facts to differentiate these two from more abstract mathematical geneticists, one of whom once suggested "considering" a population in which a recessive lethal could assume any frequency from 0 to 1.00! Paraphrasing the woman in the fast-food ad: Where's the population?

How, then, can a population whose number is regulated by food, space, and competitors be said (as the quotations cited above claim) to favor Haldane's conclusions over those of Feller? It seems to me that there may be two explanations: (1) unnecessary (and unthinking) adherence to the convention of assigning the value 1.00 to maximum fitness, and (2) the acceptance of a bookkeeper's view of life that leads to a (spuriously) precise accounting of the fate of each member of a population. By assigning a fitness value of 1.01 to the favored individual (as opposed to 1.00 for the remaining ones) and by adhering to the regulated population size as emphasized by Kimura and Ohta, I conclude that each increase in the number of favored individuals is exactly matched by the elimination of disfavored ones. It is the unreasonable loss (and then an unacknowledged recovery) of many individuals postulated under Haldane's calculation that I regard as artificial.

Kimura and Crow (1969) also cite favorably Haldane's suggestion

Dilemmas and Options

that 10% of all zygotes arising in a population might be set aside for evolutionary change; this amount is in addition to deaths caused by accidents, competition, and other factors. My views regarding this apportionment of deaths among various causes will be explained more fully in the following chapter. For the moment, I shall merely say that in my opinion a compartmentalization of deaths (environment, 40%; genetic, 30%; competition, 20%; and cost of evolution, 10%) within a density-regulated population is a hopeless (and useless) exercise. Individuals are lost not for this or that reason alone; they are lost—period! The loss of each individual in a density-regulated population, for whatever reason, increases the probability of survival for the remaining ones. At some point (as a rule), the surviving adult individuals actually form what we consider to be *the* population—the density-regulated population.

6

HARD AND SOFT SELECTION

Haldane (1957) opened his account of the cost of natural selection with the following paragraph:

> It is well known that breeders find difficulty in selecting simultaneously for all the qualities desired in a stock of animals or plants. This is partly due to the fact that it may be impossible to secure the desired phenotype with the genes available. But, in addition, especially in slowly breeding animals such as cattle, one cannot cull even half the females, even though only one in a hundred of them combines the various qualities desired.

Without its being intended to do so, that paragraph, in my opinion, describes the operation of most instances of natural selection. Despite the desires of dairymen for cows with qualities that have not been met, cattle do exist. They have not been threatened with extinction just because they fail to meet the mental constructs of dairy breeders. In truth, stanchions exist in dairy barns and cows will be found to fill them. This is the essence of what I have called *soft selection* (Wallace, 1968b), and which I equated later with frequency- and density-dependent and with rank-order selection (Wallace, 1975a).

Still other "layman" examples might be cited to illustrate soft selection. In the past I have cited the deanship at a college: let the current dean retire or leave for any reason whatsoever, and the position will be promptly filled once more. The position is filled not because large numbers of professors with deanlike qualities wait in the wings, ready to occupy the vacant chair, but because the academic position exists, because deans are necessary in fulfilling certain administrative duties, and, therefore, a body willing to bear the title *dean* must be found. This account does not suggest that the person named to fill the vacant position is chosen at random; on the contrary, finding a dean often necessitates a time- and energy-consuming search. Nevertheless, the overriding consideration is that the vacancy be filled.

A further example can be found in college athletics. A football team (American style) consists of eleven players, one of whom (the center) is responsible for transferring the ball to a person standing behind him. Regardless of the poor quality of the candidates for the center's position, no coach would settle for a ten-man team that lacked this player. A center will be on the team because the position exists. The filling of that position is an example of soft selection.

The football team may usefully be contrasted with a boxing team. A boxing team consists of individuals of different weights each of whom, when teams compete, must face, one-on-one, an opponent of his own weight classification. If no candidate of acceptable ability is available for a particular weight class (middleweight, for example), that position will remain vacant. The sport is a dangerous one in which decisions are based on individual bouts; most coaches would hesitate to endanger the health (perhaps the life) of an incompetent fighter. A boxing team, then, illustrates what I have called *hard selection*; under certain circumstances, a college may have no team at all because no candidate in any weight class possesses the necessary minimal qualifications.

The algebra of soft selection

In order to qualify for the designation *density dependent* or *frequency dependent*, the selection coefficient itself must be a function of gene frequency (p or q) or population size (density, for a specified area), N. Because gene frequencies can be expressed as $X/2N$ or $Y/2N$, where $X + Y = 2N$ ($2N$ = the number of alleles in a population of N individuals), most (if not all) frequency-dependent selections are also density dependent, and vice versa.

Fifty Years of Genetic Load

A hypothetical (and simplistic) example involving bears can be used to illustrate soft selection. During the winter, the bears must hibernate in lairs; otherwise they die. Two alleles (A and a) at a particular locus generate bears of three genotypes (AA, Aa, and aa) and two phenotypes (AA and Aa are aggressive, aa are submissive); dominance is complete.

Any bear can survive the winter provided that it occupies a lair. Aggressive bears, however, can displace submissive ones; the latter must then seek new lairs. The search for a vacant lair or, in the case of aggressive bears, for either a vacant lair or one occupied by a submissive animal, is always successful if appropriate lairs exist; no lair goes unoccupied unnecessarily.

The following calculations apply to a locality encompassing N bears and containing L lairs.

Case 1: ($L > N$)   This case requires little discussion because, with more lairs than bears, every bear can survive. The relative fitness of AA, Aa, and aa individuals are the same (1.00).

Case 2: [$L < (1 - q^2)N$]   If the frequencies of A and a are p and q, and if those of AA, Aa, and aa individuals are p^2, $2pq$, and q^2, the combined number of AA and Aa individuals equals $(1 - q^2)N$. With too few lairs to accommodate even the aggressive bears, the submissive ones will die. Some of the aggressive ones will die, too, but these deaths are randomly distributed among AA and Aa individuals. Hence, the situation with respect to relative fitnesses can be tabulated as follows:

Genotype	AA	Aa	aa
Frequency	p^2	$2pq$	q^2
Fitness	1	1	0 ($s = 1.00$)

Case 3: [$(1 - q^2)N < L < N$]   In this case, lairs outnumber aggressive animals; the remaining lairs, consequently, are available for occupation by submissive bears. However, the number of remaining lairs is insufficient to accommodate all the submissive animals. The proportion of aa individuals that survives can be calculated as

$$\frac{[L - (1 - q^2)N]}{q^2 N}.$$

Hard and Soft Selection

This expression can be simplified to $1 - [(N - L)/q^2N]$, which is equivalent to the standard notation, $1 - s$. Because the selection coefficient, s, corresponds to a term that includes both N and q, selection in this case is both density and frequency dependent: it is soft selection.

Reviewing the three cases, one sees that the fitness of the submissive animals, depending upon the prevailing situation with respect to the numbers of lairs provided by the environment, can equal that of the aggressive ones ($s = 0$), can be nil ($s = 1.00$), or can have an intermediate value [$s = (N - L)/q^2N$; where $(N - L) < q^2N$]. This is soft selection and, in my opinion, the commonest form of natural selection. This view is not shared by all (see Simmons and Crow, 1977:74).

The above example is an admittedly simplistic one, but, we might recall, most mathematical models in biology are simplistic, even those that appear to be complex. For example, Clarke (1973a, b) calculated the effect of mutation on population size. Wallace (1975a, 1981: 432ff.) examined these calculations with respect to hard (density- and frequency-*in*dependent) and soft (density- and frequency-dependent) selection. The results are shown in Figure 6-1.

In theory, selection can be either density dependent or density *in*dependent; similarly, it can be either frequency dependent or frequency *in*dependent. Consequently, one can represent the interactions of density and frequency by means of a two-by-two diagram as shown in the center of Figure 6-1.

The most frequently used (but admittedly inadequate; see, for example, Ayala et al., 1973) mathematical model describing the growth and regulation of populations of competing species (the Volterra-Lotka model) involves three variables: r_1 and r_2, the intrinsic rates of increase of species 1 and species 2; α_1 and α_2, coefficients of competition that describe the relative demands on resources that are made by members of the two species; and k_1 and k_2, the carrying capacity of the environment in terms of the numbers of individuals (or, alternatively, their biomass) of each species that the environment can support. These variables and their interrelations determine whether one species will displace the other, whether the two species will establish a stable equilibrium with both species coexisting, or whether there exists a critical point (unstable equilibrium) that determines which of the two species will displace the other.

Fifty Years of Genetic Load

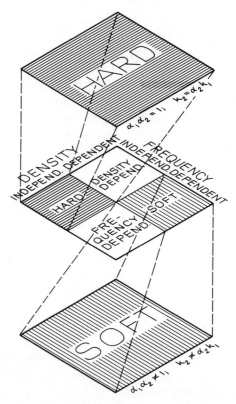

Figure 6-1. Three-tiered diagram illustrating (center) the two-by-two chart generated by frequency and density dependence (or *in*dependence); (top) the frequency- and density-*in*dependent (hard) selection that prevails, for example, if both coefficients of competition (α_1 and α_2) equal 1.00 and if the carrying capacity of the environment (k_1 and k_2) is the same for individuals of the two phenotypes; (bottom) the frequency- and density-dependent (soft) selection that prevails under virtually all other conditions. Frequency-dependent (but density-independent) and density-dependent (but frequency-independent) selection is rare or absent because frequency itself is dependent on numbers: the frequencies p and q of two alleles, A and a, can be represented as $X/2N$ and $Y/2N$, where $X + Y = 2N$, and N is the number of individuals in the population. (From Wallace, 1981, courtesy of Columbia University Press.)

Clarke (1973a, b), using a model (the Michaelis model) that is related to but not identical with the Volterra-Lotka model, considered *AA* and *Aa* individuals as being analogous to one species, and *aa* individuals as analogous to another. Applying his procedures to the Volterra-Lotka model, one can demonstrate that the relative fitnesses of these genotypic classes (the relative fitness of AA and Aa individuals equals 1.00; that of aa individuals, $1 - s$) involve both gene frequency and population size (N) unless $\alpha_2 = 1/\alpha$, and $k_2 = \alpha_2 k_1$, where $N_1 = (1 - q^2)N$, $N_2 = q^2 N$, and $N_1 + N_2 = N$.

One can envision the improbability that selection will be density— and frequency—*in*dependent by imagining the area occupied by a graph in which N_1 is the horizontal axis and N_2 the vertical one. A straight line connects k_2 on the Y-axis ($N_1 = 0$) with k_2/α_2 on the X-axis ($N_2 = 0$). A second straight line connects k_1 on the X-axis ($N_2 = 0$) with k_1/α_1 on the Y-axis ($N_1 = 0$). However, if $\alpha_2 = 1/\alpha_1$, and $k_2 = \alpha_2 k_1$, then $k_1/\alpha_1 = k_2$, and $k_2/\alpha_2 = k_1$. Hence, these two lines are superimposed. Any point lying off these superimposed lines represents conditions under which selection depends upon both population size and gene frequency (see Wallace, 1989a).

Hard and soft selection:
Contrasting consequences

The standard models used by Haldane and other mathematical geneticists in depicting the relative fitnesses of individuals of different genotypes were based on hard selection. In the case of a single locus with two alleles, individuals of genotypes AA, Aa, and aa were assigned fitnesses 1, 1, and $1 - s$; or 1, $1 - hs$, and $1 - s$; or $1 - S$, 1, and $1 - T$. The selection coefficients—s, hs, S, and T—do not involve p and q, the frequencies of the two alleles; nor do they involve N, the number of individuals. The patterns of selection are said to remain unchanged regardless of the size of the population, its density, or the relative proportions of individuals in a given area. The concept of genetic load and the calculated sizes of genetic loads were based entirely on such models of hard selection (see Figure 6-2).

One consequence of this emphasis on hard selection has been a temptation among workers to compartmentalize or classify the deaths that occur among the young zygotes of nearly any natural population. Because the causes of these deaths are seen as independent events, an increase in any one compartment (say, genetic load) carries with it the danger of extinction of the population. This danger can be met, apparently, only by an increased reproductive effort. This view of popula-

Fifty Years of Genetic Load

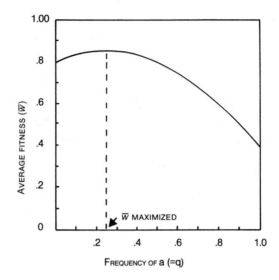

Figure 6-2. The relationship between $\bar{w}$, the average fitness of the population, and q, the frequency of a, when the relative fitnesses of AA, Aa, and aa individuals are 0.80, 1.00, and 0.40, respectively. Note (1) that the maximum $\bar{w}$ is attained when the frequency of $a = 0.2/(0.2 + 0.6)$, or 0.25, and (2) that the maximum value attained by $\bar{w} = [1 - (0.2 \times 0.6)/(0.2 + 0.6)]$, or 0.85. (From Wallace, 1987d, courtesy of *Journal of Heredity*, copyright 1987 by the American Genetic Association.)

tion dynamics was clear even in Muller's (1950) paper on the mutational load of human populations. Having calculated that mankind's genetic load is 0.20, and therefore that the approximately 2.3 children per couple in Western societies represent the surviving residue of about three zygotes, Muller declared that doubling the genetic load (thus reducing the probability of surviving for each zygote to 0.60) would threaten the human race with extinction. The basis for this declaration was simply that $3 \times 0.60 = 1.80$ children per couple; that is, children would no longer numerically replace their parents. Overlooked by Muller was an obvious fact: the reproductive behavior of human beings is based on the number of surviving children that are clamoring underfoot or out working the family farm, not on an initial number of zygotes, a number that is both unknown and unknowable.

Haldane (1953) presented a graphic model for population regulation (shown in modified form in Figure 6-3) that is based upon the simple fact that (ignoring males) a population possesses a stable size

Hard and Soft Selection

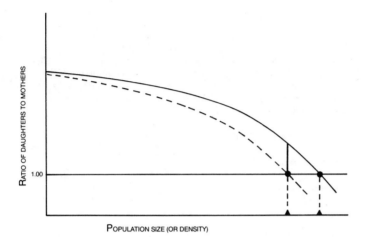

Figure 6-3. The relationship between the ratio of the number of adult, reproducing daughters of one generation to that of their mothers in the previous generation, and population size (or density, if area is assumed to be constant). Note that the population grows in size if the mothers of the current generation are more numerous than those of the preceding one. Note, too, that as the population grows in size, the daughter-to-mother ratio approaches 1.00 (where population size stabilizes) because either the mortality of progeny individuals increases or the mothers' average fertility decreases. The exposure of the population to a new cause of death (dashed line) may merely cause the population to shrink until the lessening of former deaths counterbalances the impact of the new one (vertical bar).

when, on the average, each mother leaves one adult, reproducing daughter. J. R. G. Turner and M. H. Williamson (1968) reemphasized what is implicit in Haldane's model (and was well known to both A. R. Wallace and Charles Darwin): the excess progeny of any population must be eliminated. They must die because even migration is only a temporary solution to overcrowding. In Turner and Williamson's terminology, all zygotes above an average of two per couple constitute the "ecological load" of a population or species; this load exists whatever the population's genetic composition. Genetic load, according to Turner and Williamson, represents a portion of the larger ecological load. Under this view, the independent compartments that are stressed under most genetic load calculations vanish. Malaria, for example, is an environmental cause of death in human populations lacking the allele (A^s) for sickling hemoglobin; on the other hand, it constitutes the bulk of all genetic deaths (of the segregational load) in a population

Table 6-1. The relative responses in terms of resistance (R) and susceptibility (S) of two selected lines of chickens to various disease-related or stressful challenges. Line *HA* was selected for high antibody response to injections of sheep red blood cells; line *LA* was selected for low response. When tested, more than 90% of the *HA* birds were homozygous for the B^{21} haplotype of the B-alloantigen complex; conversely, more than 90% of the *LA* birds were homozygous for the B^{13} haplotype. (Data from Gross et al., 1980; see Wallace, 1959b, 1981:247.)

Challenge	HA	LA
Eimeria necatrix	R	S
Escherichia coli	S	R
Feather mites	R	S
Mycoplasma gallisepticum	R	S
Newcastle disease	R	S
Splenomeglia virus	R	S
Staphylococcus aureus	S	R

that is polymorphic for both normal and sickling hemoglobins. (See Table 6-1 for a further example of genetic variation that, if misinterpreted, would merely suggest that chickens are subject to various environmental stresses; when fully understood, however, these stresses must be included under "genetic load.") Haldane's (1957) assertion that perhaps no more than 10% of the reproductive effort of a species can be expended on gene substitution implies that the fate of this proportion of zygotes is independent of the fates of all other zygotes. Actually, even if these zygotes are not eliminated during gene substitution, they will be eliminated nevertheless, but for other reasons.

Haldane's model, illustrated in Figure 6-3, does not allow for genetic differences between individual members of a population. As in much of the earlier ecological literature (and implied, for example, in the Volterra-Lotka model), each individual is considered to be an accurate copy of every other individual in the population. Except where age distributions are intentionally included in analyses, ecologists often look upon individuals as interchangeable units.

The recognition that a population consists of individuals of diverse genotypes leads to a more complex version of Haldane's model, as shown in Figure 6-4, which focuses on the point at which the number of surviving daughters is close to or equal to the number of mothers in the previous generation—that is, on the equilibrium population size.

Hard and Soft Selection

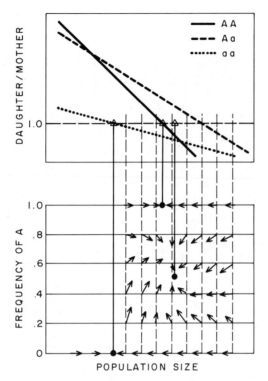

Figure 6-4. A population consists of many sorts of individuals (shown here as *AA*, *Aa*, and *aa*) whose trajectories near the point of stable population size (see Figure 6-3) need not be identical. Note that monomorphic populations (*AA* and *aa*) have stable population sizes at which daughters replace their mothers in succeeding generations. Notice, too, that a polymorphic population containing both *A* and *a* will stabilize when the daughter-to-mother ratio equals 1.00; at that point *Aa* mothers leave more than one daughter, on the average, while *AA* and *aa* mothers leave fewer than one. Despite having gained a genetic load by virtue of its being polymorphic, the population containing both alleles may be larger. The lower diagram reveals that the polymorphic population can be in a nonequilibrium state with respect to both size and gene frequency; this point recurs in Figure 10-7. (From Wallace, 1981, courtesy of Columbia University Press.)

As the population approaches its equilibrium size, not all categories of individuals (for example, those with genotypes *AA*, *Aa*, and *aa* as shown in Figure 6-4) will follow identical trajectories. In the figure, one sees that both monomorphic populations (all *AA* or all *aa*) as well as the polymorphic one containing both alleles can exist with no difficulty. Relative fitnesses such as those listed in Table 6-2, many

Table 6-2. The relative adaptive values of various genotypes in experimental populations of *Drosophila pseudoobscura* monomorphic or polymorphic for the *ST*, *AR*, and *CH* gene arrangements, from Piñon Flats, California. (After Wright and Dobzhansky, 1946.)

Gene arrangements in populations	Genotype						
	ST/ST	ST/AR	ST/CH	AR/AR	AR/CH	CH/CH	$\overline{W}$
ST	1.000	—	—	—	—	—	1.000
AR	—	—	—	1.000	—	—	1.000
CH	—	—	—	—	—	1.000	1.000
ST + AR	0.81	1.00	—	0.50	—	—	0.862
Adjusted	0.94	1.16	—	0.58	—	—	1.000
ST + CH	0.85	—	1.00	—	—	0.58	0.889
Adjusted	0.96	—	1.12	—	—	0.65	1.000
AR + CH	—	—	—	0.86	1.00	0.48	0.890
Adjusted	—	—	—	0.97	1.12	0.54	1.000
ST + AR + CH	0.43	1.30	1.00	0.05	0.71	0.21	0.799
Adjusted	0.54	1.63	1.25	0.06	0.89	0.26	1.000

persons forget, are relevant for the population at stable size and might change considerably if estimates could be made for nonequilibrium conditions. The lower diagram in Figure 6-4 shows that whereas monomorphic populations can be displaced from equilibrium only with respect to population size, polymorphic ones can be displaced with respect to size, gene frequency, or both.*

Unit spaces

Still another model illustrating soft selection can be described here, although its full significance, I suspect, will become apparent only in a subsequent chapter (Chapter 8). In this model (which is most easily visualized for plants), we imagine that the environment is divided into small areas, each of which is capable of supporting only one individual of a given age or size. Seeds fall into these spaces at random (for example, according to the Poisson distribution): some spaces remain empty, some are occupied by one seed, others by two or more seeds.

The empty spaces will remain empty. Spaces occupied by single seedlings will support these individual plants whatever their nature (other than intrinsic lethality or near lethality—that is, disabilities that are frequency and density *in*dependent); later, of course, when the seedlings outgrow their initial spaces, as in the case of large bushes or trees, one must imagine larger spaces that may now once again include competing individuals. This unit-space model illustrates soft selection because, spared mistreatment by a larger, stronger competitor, even an initially delicate individual (a late germinator, for example) can survive to reproductive age.

In those cases containing two or more young seedlings, there is a

---

*Whereas Figure 6-3 shows the relationship between the daughter-to-mother ratio (D/M) and population size as a curved line intersecting the line D/M = 1 at a point, Figure 6-4 reveals that the point is in reality a vertical line through which numerous trajectories pass, some above D/M = 1, some below. The vertical line, in turn, proves to be the end-on view of a plane that encompasses a large number of microenvironments and background genotypes, as well as their many interactions. The "point" of intersection shown in Figure 6-3, then, is in reality a surface possessing both height and width within which, simply because it represents an equilibrium "point," virtually all selective events of evolutionary interest transpire. These topics recur in Chapter 10 (p. 149).

Fifty Years of Genetic Load

Figure 6-5. Relative sizes of silver maple (*Acer saccharinum*) seedlings that sprouted from fruit that fell into small pots placed beneath the parental tree: some pots contained no seeds, others had one, and a few had two. Three singlets are shown at the top, three doublets at the bottom. Note that one member of each competing pair is comparable in size with the singlets that grew without competition. The other member of each competing pair, on the other hand, is moribund.

shortage of resources (sunlight, nutrients, or water) that results in what is generally recognized as "competition." Figure 6-5 illustrates the outcome in the case of an experiment involving silver maple (*Acer saccharinum*) seedlings. In each of the three cases (lower row) in which two fruits fell into the same small pot (the unit space), the outcome was quickly decided: one seedling shaded the other, which then became moribund. The larger seedling of these pairs suffered no harm by virtue of the "competition"; each of the three thriving individuals of these competing pairs is comparable in size to the larger of those seedlings that grew without nearby competitors (top row).

The outcome of the unit-space model is that the number of surviving individuals equals the number of spaces containing one or more seedlings. This model was first applied to adult *Drosophila melanogaster* flies emerging within 2.5 mm × 100 mm shell vials containing a rich but rather watery medium (Wallace, 1968b). A chi-square test for homogeneity revealed that the total number of pupae formed in different vials was extremely heterogeneous, whereas the number from which surviving adults had emerged was essentially homogeneous—so much so that the probability of seeing deviations as small as or smaller than those observed among different vials was less than 0.05.

A case study in soft selection: the *corky* gene in New World cotton

About forty years ago, S. G. Stephens (1946, 1950) described the characteristics and geographical distributions of recessive, complementary dwarfing genes in two New World cotton species, *Gossypium hirsutum* (X) and *G. barbadense* (Y). The locus involved was designated *Corky* (Ck). In each species a recessive allele, *corky*, was found; these alleles can be identified as ck_x and ck_y for convenience. Homozygous ck_x/ck_x and ck_y/ck_y plants do not differ in any obvious way from normal plants (Ck/Ck, Ck/ck_x, or Ck/ck_y) of the corresponding species. The effect of the recessive alleles is seen only in interspecific hybrids: ck_x/ck_y hybrids are dwarf plants that are heavily encrusted with corklike bark. Such "corky" hybrids differ considerably from other hybrid individuals carrying one or two Ck alleles; the latter exhibit considerable hybrid vigor, so much so that cotton growers have frequently considered utilizing heterosis in cotton production much as midwestern farmers have used it in developing hybrid strains of corn (see Figure 6-6).

Although the ck_x and ck_y alleles are found here and there throughout the geographic ranges of both species of cotton, they are extremely common in areas where the two species' ranges overlap and the plants of both species live sympatrically. Stephens, having noted the correla-

Fifty Years of Genetic Load

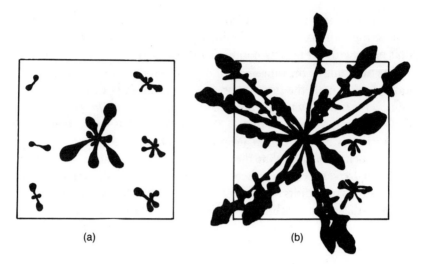

Figure 6-6. Complementary recessive lethal genes (l_c and l_t) in the closely related species *Crepis capillaris* and *C. tectorum*. The recessive allele has no effect on viability in either species, even when homozygous (l_c/l_c or l_t/l_t). The combination l_c/l_t in interspecific F_1 hybrids, however, severely retards the development and growth of its carriers. As in the case of the *corky* lethal alleles of cotton discussed in the text, soft selection may favor genes that cause the death of already sterile interspecific F_1 hybrids. (a) One viable interspecific F_1 hybrid surrounded by six lethal ones (l_c/l_t). (b) The same plants one month later. Only two, severely stunted l_c/l_t plants remain; the viable hybrid, in contrast, exhibits luxuriant growth. (After Hollingshead, 1930.)

tion between sympatry and the high frequency of the *ck* alleles, suggested that ancient cotton growers were responsible; the alleles ck_x and ck_y revealed which plants were of hybrid origin, thus warning the growers not to collect the seeds they bore; seeds of F_1 hybrid plants produce, according to Stephens, a medley of unthrifty types. Stephens casually mentioned the matter of competition for resources between hybrid and nonhybrid individuals but concluded that the recessive alleles, ck_x and ck_y, must be neutral or even advantageous within their corresponding species if their existence is to be understood. No obvious (Darwinian) selective advantage can seemingly be ascribed to an allele that merely stunts an already effectively sterile hybrid.

The standard account of selection (see p. 9) stresses the removal of deleterious alleles from populations; the change in gene frequency for rare alleles is negative—approximately $-sq^2$. Soft selection (or rank-order selection) casts matters in a quite different light, however. Be-

Hard and Soft Selection

cause culling begins at the low end of the fitness distribution and then proceeds toward the upper end, the presence of extremely vigorous hybrid individuals in a population of competing zygotes threatens the survival of even the most vigorous members of either species. An analogous problem faced the primitive cotton growers in their choice of seed for subsequent planting. Thus Stephens ascribed the selection for the recessive *corky* alleles to human intelligence: these alleles destroyed the apparent (but false) desirability of interspecific hybrids.

An alternative explanation to human intervention exists, however (Figure 6-7): any plant that reduces the number of vigorous hybrids among its progeny automatically increases the chances for the survival of its legitimate, pure-species offspring. The figure shows the types of offspring, pure species and hybrid, produced by plants of species X after exposure to a thoroughly mixed cloud of pollen grains of both species X and Y. Unlike pollen grains, most seeds, because of their weight, tend to fall and remain near their mother plant, thus giving rise to competition among what are, for the most part, half sibs. The more vigorous (including rapid germination as a possible component of "vigor") individuals among such progenies tend to survive; by far the most vigorous cotton plants, however, are the interspecific hybrids.

Figure 6-7 emphasizes that ck_x/ck_x (or, viewing matters symmetrically, ck_y/ck_y) plants, by producing a number of dwarf (ck_x/ck_y) rather than vigorous hybrid offspring, provide a greater assurance of survival to their pure-species progeny than do their Ck/Ck and Ck/ck_x (or Ck/ck_y) counterparts. This assurance (or lack of it) can be translated into relative fitnesses that apply to *progenies* of parental plants of different genotypes (Ck/Ck, Ck/ck_x, and ck_x/ck_x, or, for the other species, Ck/Ck, Ck/ck_y, and ck_y/ck_y) (see Figure 6-8). Further, noting that T is smaller than S in the figure, it can be shown that there is no intermediate equilibrium frequency: the allele ck_x (or ck_y) will tend to displace Ck in areas within which interspecific fertilizations occur. An essential feature of this pattern of selection is the assignment of fitness values to individuals of different genotypes according to the survival of the offspring they produce; this aspect of selection was emphasized earlier in this book (Chapter 2), where I claimed that the shortcomings of progeny could be ascribed to the fertility of the parents.

Many consequences follow from the pattern of selection illustrated in Figures 6-7 and 6-8, where the *corky* gene has served as an example.

Fifty Years of Genetic Load

Plants of Species X			Pollen				Fitness
			Species X		Species Y		
			Ck	ck_x	Ck	ck_y	
Genotype	Freq	Ova	p	q	p	q	X
Ck Ck	p^2	Ck	CkCk	$Ckck_x$	CkCk	$Ckck_y$	1-S
Ck ck_x	2pq	Ck	CkCk	$Ckck_x$	CkCk	$Ckck_y$	1-T
		ck_x	$Ckck_x$	$ck_x ck_x$	$Ckck_x$	$ck_x ck_y$	
ck_x ck_x	q^2	ck_x	$Ckck_x$	$ck_x ck_x$	$Ckck_x$	$ck_x ck_y$	1

Figure 6-7. The selective advantage that resides with ck_x/ck_x (or, to speak of the other species, ck_y/ck_y) individuals within a geographic area occupied by both species, X and Y. Pollen of the two species, it is assumed, is more thoroughly mixed than are the seeds of individual plants. Thus ck_x/ck_x maternal plants are assigned highest (relative) fitness because by producing *corky* hybrid progeny they lessen the local competition facing their legitimate offspring. In contrast, of the progeny of *Ck/Ck* plants, approximately one-half are vigorously competing hybrid individuals capable of crowding out all but the luckiest of their pure-species sibs. (From Wallace, 1988, courtesy of *Journal of Heredity*, copyright 1988 by the American Genetic Association.)

The pattern can be extended to dominant rather than recessive alleles, and to two-locus complementary lethals that involve either dominant or recessive alleles. This is not the place to provide extended analyses of these additional possibilities (see Wallace, 1988). Suffice it to say that the situations involving plants and animals are clearly different because of the essentially multiple inseminations (including interspecific hybridizations) sometimes endured by individual plants in contrast to the single inseminations that are characteristic of many, if not most, higher animals. Thus, genetic mechanisms that reduce the competitive nature of the sterile hybrid members of a given plant's progeny can be favored, as illustrated in Figures 6-7 and 6-8; such mechanisms would seemingly be favored only on an altruistic-like mode in the case of singly inseminated animals (see Coyne, 1974).

The case study involving the *corky* gene in cotton can be continued by making two additional points. First, there is no need to assume, as Stephens did, that the recessive allele (ck_x or ck_y) is neutral or advantageous in the standard (Darwinian) sense. An examination of Figure 6-8 reveals that a selective disadvantage (*s*) could be assigned to the

Hard and Soft Selection

	$CkCk$ p^2	$Ckck_x$ $2pq$	ck_xck_x q^2
	1-S S	1-T T	
$CkCk$	░░░░░░	$CkCk$ ------- $Ckck_x$	$Ckck_x$
$Ckck_x$	░░░░░░	$Ckck_x$ ------- ck_xck_x	ck_xck_x

Figure 6-8. A diagram adopted from Figure 6-7 that simplifies the calculation of gene frequency changes (Ck and ck_x) within species X in the presence of species Y. (Because the situation is viewed as being symmetrical, similar changes in the frequencies of Ck and ck_y would occur within species Y in the presence of species X.) Calculations based on this diagram reveal that Ck and ck_x will establish equilibrium frequencies only if T is greater than S; ck_x will increase in frequency within the geographic zone of overlap, supplanting Ck, unless ck-bearing individuals (ck_x/ck_x or Ck/ck_x) exhibit a lowered Darwinian fitness. The model presented in Figure 6-7 specifies that T is smaller than S. (From Wallace, 1988, courtesy of *Journal of Heredity*, copyright 1988 by the American Genetic Association.)

genotype ck_x/ck_x—both to the homozygous parent and to homozygous ck_x/ck_x offspring—without necessarily canceling the advantage gained by the dwarfing of the otherwise vigorous hybrid half sibs in individual progenies. As the frequency of ck in the population increased, however, the possible disadvantage, s, of ck/ck individuals would become increasingly important, and, as a result, the frequency of ck could reach an intermediate equilibrium value where the contrasting selective forces are equal.

The second point concerns a more generalized account of rank-order selection (selection that I equate with soft selection) and its possible consequences in the *corky*-like response of plant species to the production of vigorous, though sterile, hybrid individuals. Six diagrams (a–f) are shown in Figure 6-9. At the left (diagrams a, c, and e) are illustrated the relative "fitnesses" of two species (A and B) and their A/B hybrid. The changing relative fitnesses of A and B result from a clinical environmental change (not illustrated) extending from left to

Fifty Years of Genetic Load

Figure 6-9. The relative fitnesses (left columns) of the members of two plant species (*A* and *B*) and of their hybrid, and (right column) the distributions of *A*, *B*, and hybrid individuals that result from the corresponding fitness relationships. Throughout, a left–right environmental gradient (not shown) causes the fitnesses of *A* and *B* to vary inversely from one end of the gradient to the other. The top diagrams (a and b) show that the species would exhibit an abrupt turnover in numbers at the point where the standings of their relative fitnesses are reversed. The middle pair (c and d) show that the frequent formation of vigorous F_1 hybrids, even if the hybrids were sterile, would lead to a zone in which many (perhaps most) surviving individuals would be of hybrid origin. The bottom pair (e and f) illustrate the elimination of the hybrid zone by the occurrence of mutations (shown here as complementary lethals) that incapacitate the hybrid plants but only have minor effects on members of the individual species. The *Corky* gene in cotton illustrates the latter case; it is discussed in the text (see Figure 6-6).

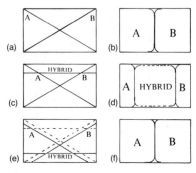

right in each diagram. At the right, diagrams b, d, and f illustrate the actual proportions of individuals—*A*, *B*, or hybrid—that would be expected, given the relative fitnesses illustrated at the left.

The top pair of diagrams (a and b) show that if their relative fitnesses vary inversely over the environmental gradient as shown in a, the two species will abut one another along a reasonably sharp boundary; the species with the superior fitness at any locality displaces the other. If either species were to be removed from the picture, the other would probably extend its range (see Figure 6-10).

The second pair of diagrams (c and d) shows the consequences of hybridization and the production of vigorous (though sterile) F_1 interspecific hybrids. The vigor of the hybrids is such that these individuals exceed members of either species (progeny by progeny for the most part) in competitive ability in all but the extreme portions of the environmental gradient. The consequence (diagram d) is a zone of considerable size in which hybrid individuals predominate, although not necessarily to the exclusion of the parental species. The presence of the latter is, of course, necessary for the continued production of hybrids.

The last pair of diagrams (e and f) illustrate the consequences following the origin of *corky*-like mutations in the two species. These

Hard and Soft Selection

Figure 6-10. The abutment of three species of *Drosophila* in central Texas. Dobzhansky (Dobzhansky and Epling, 1944:15) reports that "the very extensive collecting of Patterson and his colleagues in Texas shows that the eastern boundary of [*D. pseudoobscura*] in that state is close to the line Wichita Falls–Plano–Fort Worth–Arlington–Florence–Georgetown–Aldrich near Austin–San Antonio–Three Rivers–Alice–Falfurrias [solid circles and curved line]. Here it comes in contact with the westernmost extensions of the ranges of two species of the *affinis* group, *D. affinis* and *D. algonquin*." The western border of the geographic range of *D. pseudoobscura* is the Pacific Coast of the United States. Within Austin, Texas, *D. pseudoobscura* is found in the hill region of the western suburbs but not in the lower-lying eastern part of the city.

mutations lower the competitive nature of the hybrid individuals considerably; they may even have deleterious effects (as shown) on their carriers within species *A* and *B*. Nevertheless, unless they severely incapacitate their intraspecific carriers, their effect is to remove the zone within which hybrids would otherwise predominate, and to extend the range of each species until the two abut as in diagrams a and b. The frequencies of the *corky* alleles would be high in the center of the diagram but could be lower at the margins, where hybrids pose no problem.

A final sentence may reassure many readers: At no point in this discussion of *corky* and *corky*-like mutations has group selection been invoked; the operative mechanisms have been soft (i.e., rank-order) selection, the nonrandom spatial distribution of the seeds of individual plants, and the referral of fitness values from individual progenies to their parental plants. The essential problem concerns the welfare of one's progeny. Birds often face similar problems in seeking suitable sites for building nests; parental birds that consistently choose poorly may prove to be effectively sterile.

Personal comments

To appreciate the points made in this chapter, we shall return once more to the calculations of Kimura and Crow (1964), in which it appeared that eight alleles per locus could be maintained in a population of 10,000 individuals if all heterozygotes had a 1% selective advantage over all homozygotes. If this were the case, the load imposed by a single locus would be $\frac{1}{8} \times 0.01$, or 0.00125. The average fitness for a single locus would be 0.99875, but if 5000 loci exhibited such polymorphisms, the average fitness would be $(0.99875)^{5000}$, or 0.002.

The average fitness (0.002) calculated here is relative to that of an individual heterozygous at all 5000 loci. How frequent would such an ideal individual be in a population? The probability of being heterozygous at any one locus equals 0.875; the probability of being heterozygous at 5000 loci equals $(0.875)^{5000}$, or 10^{-290}. In brief, there are no totally heterozygous individuals against which to compare the population average! If one uses instead the binomial distribution for computing means and variances, one finds that the expected (average) number of heterozygous loci would equal 0.875×5000, or 4375. The standard deviation for the distribution about this mean is 25; thus, 95% of the population would be heterozygous for some number between 4325 and 4425 loci (see Sved et al., 1967). What are essentially the best and the poorest members of the population (with respect to fitness) differ by homozygosity at 100 loci. The fitness of the poorest relative to that

Hard and Soft Selection

of the best equals $(0.99)^{100}$, or 0.37. The value 0.002 simply does not appear in these calculations. Nor do these newer calculations reveal the entire story: If only the lower $2\frac{1}{2}$% of the population were to survive some perverse environmental catastrophe, the best of these would now assume fitness 1.00, not 0.37 as calculated in the previous sentence.

In reassessing the calculations initially performed by Kimura and Crow (1964), Kimura and Ohta (1971:58) consider the possibility that selection may be based on competition between those genotypes that are actually present in a population; this has led them to compute the "most probable largest number of heterozygous loci," a calculation that lessens the original segregational load (0.998) considerably. This modified calculation is still faulty, however; the survival of a single individual (no matter how fit) need not greatly lower the probability of survival of all other members of a population. This point became clear in discussing Haldane's (1957) computation of the cost of natural selection: in a population of 100,000,001 individuals, one need not assume that the single individual possessing a 1% selective advantage will in some manner cause the death of 1,000,000 individuals of lesser fitness; the selectively favored individual would not even encounter that many others.

I believe that those who are accustomed to performing calculations based on hard selection (i.e., density- and frequency-*in*dependent selection) also tend to view survival from the wrong point of view. They view it from the top, not from the bottom. For example, Kimura and Ohta (1971:155) are doubtful that, under natural selection, "the total number of heterozygous loci for each individual is counted and the individual is culled if the number of heterozygous loci is less than a certain critical number." Doubt stems from their disbelief that Mother Nature can count. Their statement is intended to counter the argument of those who believe in truncation selection; namely, that there is a cutoff point that separates those individuals who fail to survive (fitness nil) from those who meet the requirements for survival and reproduction (fitness 1.00). Kimura and Ohta criticize specifically those who would have the number of heterozygous loci serve to determine that cutoff point.

Truncation models as generally proposed err in two ways. One, a trivial one, is that truncation selection based on phenotype (other than

Fifty Years of Genetic Load

fitness itself) does not eliminate natural selection; it does not divide the population, as its proponents would have us believe, into those individuals with zero fitness and a remainder with fitness equal to 1.00. Usually, the reproductive fitness of individuals exhibiting extreme phenotypes declines relative to those whose phenotypes are closer to the average (and, hence, closer to the cutoff point); the fitnesses of the individuals remaining after culling still vary.

A second, more serious, error is the view of life that truncation selection engenders; namely, that individuals above a certain point are assured of survival. Survival cannot be guaranteed! Whatever the genotypes of its members, a population must be reduced in size to the carrying capacity of the environment. Natural selection proceeds in a manner precisely the reverse of that implied by truncation selection. Culling starts at the bottom. With each individual that is culled, the probability of surviving and reproducing increases for the remaining ones. No account need be kept of the reasons why various individuals are removed from the population—no tabulation of genetic deaths, ill-nourishment, or accidents. One can imagine the course of events in either of two (related) ways: the environment has "space" for a given number of survivors, and in the end these spaces will be filled (à la the hibernating bears and dairy cattle that were mentioned earlier in this chapter), or, as more and more individuals are culled, the probability of survival for the remaining individuals increases to a point where they actually become surviving and reproducing adults. This might be thought of as forming a hyperbolic curve where N_t, the number of individuals existing at time t, times p_t, the probability of survival $(= k/N_t)$ at time t, equals k, the carrying capacity of the environment. This view allows for the carrying capacity of the environment to change in response to the nature or characteristics of the surviving individuals themselves. It also allows for both genetic and phenotypic differences among the successful survivors.* As Kimura and Ohta

---

*C. E. Blohowiac and I emphasized that in attempting to understand the fate of recessive lethals in populations of *Drosophila melanogaster*, the *average* effect of these lethals on the relative fitnesses of their heterozygous carriers may not be particularly useful information (Wallace and Blohowiac, 1985a, b). More important by far is information regarding the proportion of recessive lethals among individuals that are most likely to survive and reproduce. The proportions of lethal chromosomes among heterozygous (see Figure 3-1) chromosome combinations exhibiting the highest egg-to-adult viabilities proved to be as high as (in one case clearly higher than) the averages in

emphasize, there is no counting mechanism; there are no means by which numbers of heterozygous loci are counted. If, however, high levels of heterozygosity are selectively favored, individuals carrying appropriate genotypes will be concentrated among the final survivors—that is, among those individuals that constitute the reproducing portion of that generation of zygotes.

A final comment concerns the subtle differences in phenotype that, at times, may have tremendous effects on fitness. The late I. M. Lerner, for example, told me he was greatly impressed by the consistent genetic differences between his chicken flocks overall and the individuals picked by his research associate to form the breeding stock for the next generation. The excess of blood-group heterozygotes among those chosen for breeding purposes had a profound influence on his concept of the genetic structure of populations. Had a novice picked the breeding birds in Lerner's flock, there might have been no bias in favor of blood-group heterozygotes; only a practiced eye could detect, not heterozygous loci, but rather those physical characteristics known to be associated with breeding success. The different consequences that would have followed from Lerner's use of a novice technician would have illustrated what Kimura (1983:271) refers to as the "Dykhuizen-Hartl" effect: alleles that respond to selection in one environment (often stressful, but not necessarily so) can be neutral (or nearly so) under other circumstances. Precisely so! And, indeed, soft selection abounds with examples of the Dykhuizen-Hartl effect—even among hibernating bears.

---

the various experimental populations. Haldane (1932:177) first made this important point: intense selection selects for the more variable population, not necessarily for the one with the higher mean. Herein lies, I suspect, the generally observed poor heritability of "normal" viability: As long as natural selection tends to preserve individuals at the upper tail of a viability ($\simeq$ fitness) distribution, neither the position of the mean nor the wretchedness of the lower extreme is of any importance for the genetics of the population (see Wallace, 1989b).

7

PERSISTENCE: AN IMPORTANT COMPONENT OF POPULATION FITNESS

Population biologists speak rather glibly at times of population fitness. Population geneticists, of course, speak often of $\overline{W}$, the average Darwinian fitness of a population. Some, and I believe Sewall Wright was among them, think that the fitnesses of two populations can be compared by means of their $\overline{W}$'s. If such a comparison is made, of course, it must rest on numerous, usually implicit, assumptions. The truth is that $\overline{W}$ is an average of relative values—values that are relative to one another *within* a population. Only an assumption of virtual (genetic) identity of the populations being compared can allow one to utilize $\overline{W}$ as a basis for interpopulation comparisons.

In what way did the comparison of the irradiated populations of *Drosophila melanogaster* differ from a mere comparison of $\overline{W}$'s? The fitnesses of wild-type flies that were measured within the F_3 test cultures of the *Cy L* or *Cy L–Pm* test procedures were measures of (1) egg-to-adult viability and (2) developmental rate of wild-type flies relative to those of *Cy L/+* or *Cy L/Pm* flies that served as *external* standards for comparison. It was assumed in these tests that the wild-type chromosomes of one population would not produce *Cy L/+* flies that systematically differed in viability from the *Cy L/+* flies carrying wild-type chromosomes of another population. At least for populations whose origins involved the same strains of flies, this assumption was correct; the use of the *Cy L–Pm* procedure provided the necessary supporting evidence. The lack of bias is true, however, for averages based on large numbers of observations; it is not necessarily true that individual chromosomes do not exhibit special interactions with *Cy L* or *Pm* chromosomes. Such interactions have, as I mentioned earlier, been demonstrated (Wallace, 1963a): wild-type chromosomes that produce $M_1/+$ (or $M_2/+$) flies of poor viability do not produce $M_2/+$ (or $M_1/+$) flies of correspondingly poor viability (where M_1 and M_2 refer to two different genetically marked chromosomes carrying inverted gene arrangements). This acknowledgment of special interactions between wild-type and genetically marked chromosomes even when the former are obtained from a single population should be contrasted with the assumption made by Mukai et al. (1972) that the survival or viability of *Cy L*/+ flies is a constant when the wild-type chromosomes are derived from the same population. That assumption is demonstrably wrong.

The comparisons of the irradiated populations were not accomplished by comparing $\overline{W}$'s obtained through theoretical calculations; on the contrary, they were accomplished by subjecting developing flies to some degree of stress and measuring their ability to survive and develop. Survival and developmental rate, however, are not identical with fitness—neither with intrapopulation Darwinian fitness nor with the still-undefined "population" fitness.

In his article "Adaptedness and Fitness," Dobzhansky (1968) attempted to disentangle the adaptedness (or fitness) of populations and Darwinian or relative fitnesses of the individual members of populations. According to Dobzhansky, the former, in theory (although not readily in practice), can be assigned an absolute value; Darwinian fitness of genes and combinations of genes within populations cannot be expressed in absolute terms, however, because these fitnesses are relative to those of other genes and gene combinations in the population.

One measure of population fitness mentioned specifically by Dobzhansky (1968) is r, the intrinsic rate of increase for a population. According to studies he cited, populations of *Drosophila pseudoobscura* polymorphic for chromosomal inversions exhibited higher intrinsic rates of increase than monomorphic ones; furthermore, laboratory populations underwent increases in r with time, thus revealing their continued adaptation to laboratory conditions and providing evidence of an increasing fitness.

Different workers have used different measures of population fitness (see Wallace, 1968a and 1981, for reviews). H. L. Carson (1957, 1958, 1961), for example, used the biomass of populations and the numbers of individuals (population size) as measures of population fitness. F. J. Ayala (1966, 1968) also used population size for this purpose, but he extended his studies to include productivity (a rate) as well. Productivity, of course, is closely related to innate capacity for increase.

Interspecific competition has been used to study the fitnesses of populations (Ayala, 1969; Blaylock and Shugart, 1972). D. L. Hartl and Hans E. Jungen (1979; Jungen and Hartl, 1979) used compound-autosome strains of *Drosophila melanogaster* as competitors for wild-type strains in order to reduce the ecological differences between competing flies (compound-autosome strains produce no viable offspring when mated with normal, wild-type flies).

A sequence of positive correlations can be detected among the many empirical studies of population fitness. Results obtained by the *Cy L* and *Cy L–Pm* procedures in studying irradiated populations of *Drosophila melanogaster* are supported by population size and productivity tests of irradiated populations by Ayala (1966, 1968) and interspecific competition tests by B. G. Blaylock and H. H. Shugart (1972). Similarly, the higher fitness ($\overline{W}$) calculated for balanced polymorphic populations of *Drosophila pseudoobscura* gains credence from the higher intrinsic rate of increase observed in the polymorphic populations, the greater competitive ability of polymorphic strains (Ayala, 1969), and the greater biomass and numbers of flies emerging per food cup observed for polymorphic laboratory populations (Dobzhansky and Pavlovsky, 1961).

The studies mentioned in the preceding paragraphs yield numerical data of the sort that Dobzhansky (1968), in his attempt to disentangle adaptedness from (relative) Darwinian fitness, claimed could—in theory—provide an absolute numerical measure of population fitness: offspring produced per unit time, number of individuals per population, grams of flies produced per gram of food, grams of flies produced per unit time, or an ability to cope with and possibly eliminate competing species. Ironically, one attempting to understand the basis for the

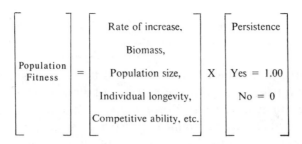

Figure 7-1. The usual measures of population fitness are related to the population's "true" (or "lasting") fitness by the matter of persistence; an extinct population has zero fitness. The text argues that some factors that make persistence more or less likely can be identified; persistence, that is, is not a matter whose evaluation lies entirely in the future, to be assessed only by hindsight.

different Darwinian fitnesses of two genotypes within a single population would study very nearly the same factors: differential survival, differential fertilities, relative abilities to find mates, relative longevities, and other factors bearing on reproductive success. The same sorts of evidence, in other words, are said to lead to absolute knowledge concerning fitness at the population level but only to an understanding of relative fitnesses *within* individual populations.

The account presented so far has omitted what in my view is the most important component of population fitness: the ability of the population to continue in existence, its *persistence* through time. Whatever this or that study may reveal about intrinsic rate of increase, competitive ability, or biomass, the extinction of a population reduces its fitness to zero, by definition—and for all time. Extinction, as the popular motto claims, is forever. In any one generation a population either continues its existence or it does not; in combination with the other factors related to population fitness, persistence provides a multiplier whose value equals either zero or one (Figure 7-1).

Unfortunately, the study of persistence requires an ability to foretell the future. J. M. Thoday (1953), who first clearly identified persistence as a component of population fitness, suggested 10^8 years as a period of time that might elapse before the relative persistences of populations were evaluated. More modest attempts at measuring persistence have been made by a number of *Drosophila* workers (personal communications), but few results seem to have been published. Such studies take the following form: What is the probability that a culture (i.e., a population) of *Drosophila* flies will exist after x generations during which the progeny flies of one generation become (i.e., are used as) the parental flies of the next? The probabilities obtained for two or more strains (through the study of many cultures for each strain) are then used in contrasting the relative persistences of these strains. In a crude sense, one might imagine a *Drosophila* stock-culture center where many strains of flies are maintained: special mutant stocks are accidentally lost more frequently than are wild-type strains; the latter, then, can be assigned a higher fitness—one based on persistence.

Laboratory studies on persistence have revealed that, because of the dangers accompanying overcrowding, virtually all aspects of a population (intrinsic rate of increase, biomass, and population size, for example) that have been used as measures of population fitness (the greater or higher, the better) can, in fact, lead to a population's demise—that is, to zero fitness. An example drawn from a "transfer" study such as that described in the preceding paragraph is provided in Table 7-1. Fourteen culture vials of *Drosophila melanogaster* were lost in the sixth transfer generation; these cultures, without exception, had had above-average numbers (130–190) of progeny two generations earlier. Gross overcrowding during the intervening (fifth) generation had led to the production of starved adults that were incapable of leaving offspring when transferred to fresh vials.

The concept of population fitness in the minds of many, I am sure, is not only elusive but also nonexistent. Some would say that a population is no more than a collection of individuals; therefore, those individuals must form the basis of any investigation into fitness. These persons would add that *population* is the reification of an abstraction. Others, while admitting that populations exist, would say that they have no fitnesses other than those of their individual members. Those who believe that fortune-telling has no place in science are inclined to say that if a population has a fitness, it must be either 1.00 or 0—the population exists or it does not exist. In contrast, I believe that one can identify at least one factor that affects a population's persistence and occupies a level above that of the population's individual members: the population's composition. This belief stems from the importance that I place on soft selection, selection that is both frequency and density dependent. Within a given environment, both the genetic composition of a population (*frequencies* of various genotypes) and its size (number

Table 7-1. The average number of *Drosophila melanogaster* produced per vial in the fifth and sixth generations by vials that in the fifth generation had received various numbers of fourth-generation progeny as parental flies. Under "zero" is listed the number of vials that yielded no progeny at all. (After Wallace, 1979.)

Progeny in parental vials (gen. 4)	No. of vials	Generation 5		Generation 6	
		No. of progeny	zero	No. of progeny	zero
50	2	116	0	51	0
70	1	72	0	145	0
90	1	8	0	12	0
110	6	25	0	46	0
130	13	22	0	44	2
150	13	16	0	20	7
170	4	3	0	3	3
190	2	3	0	0	2

or *density* of individuals) influence its fate; these matters will be discussed in the next chapter with respect to the self-culling of populations.

Personal comments

The comments made above regarding the reality of populations and the fitness of populations are not unrelated to a current topic of some controversy: At what level does natural selection act? Richard Dawkins (1976) identified the (selfish) gene as a unit of selection. Eliot Sober and R. C. Lewontin (1984) have pointed out that gene frequencies can remain constant while many individuals are selectively eliminated from a population; indeed, it is the selection of *individuals* that leads to the stability of gene frequencies: their example involves the superior fitness of heterozygotes. If one abandons density- and frequency-*in*dependent models (hard selection) and considers instead models based on soft selection, the population (species) becomes the unit of selection. Interactions between units at any level—gene, individual, or population—effectively transfer the level at which selection acts to the next higher category.

In my opinion, all proponents in this controversy are partially right and partially wrong. At the outset, of course, everyone might agree that *individuals* are the living, breathing, (among higher animals) walking, and reproducing units of populations. They carry genes; genes are not normally found strewn naked over the landscape, exposed to the environment (see, however, Davey and Reanney, 1980, for an account of microbial populations). They also constitute populations: eliminate the individual members of a population and the population itself is eliminated. Having said the above, however, one finds that still more needs saying.

If all homologous alleles lacked dominance, and if epistatic interactions among genes at different loci were lacking as well, one could reasonably insist that genes, not individuals, are the *interesting* units of selection. Of course, it is precisely dominance, heterosis, and epistatic interactions that cause individuals (as carriers of gene constellations) to be the units of selection—but units whose selection nevertheless reflects on gene frequencies as well. The interactions between and among the individual members of a population cause the population to assume properties (based on its *composition*) other than those of its individual members and, hence, to become effective units of selection in turn. The requisite interactions are those that result in frequency- and density-*de*pendent selection. Finally, of course, interactions among species endow the ecological community with properties that cannot be ascertained by a study of individual species in isolation—even by a study of every species included within the community. These species interactions impose a modicum of selection on the ecological community itself. That is, if the community composition were not correct, some species would be lost even though the types of individuals constituting these ill-fated species might correspond precisely to those of a surviving population located elsewhere.

8

SELF-CULLING AND THE PERSISTENCE OF POPULATIONS

Overcrowding was identified in the previous chapter as a threat to a population's persistence. Upon learning of a small study I was performing on the relationship between the degree of crowding and seed set in a small mustard plant (*Brassica campestris*), a senior, plant-breeder colleague remarked that I was attempting to prove what every farmer knows: if crops are seeded too densely, the growing plants are stunted and seeds do not mature. A dense stand of wheat might produce a fine lawn, but no wheat for harvesting.

The mortality of insects in laboratory cultures is often a function of the initial number of individuals in the culture. Thus, letting x be the number of survivors and N the initial number of individuals, the probability of surviving can often be written

$$\frac{x}{N} = a - bN,$$

corresponding in form to the more common equation

$$Y = a - bX.$$

Rearranging this equation leads to

$$x = N(a - bN).$$

Thus, x, the number of survivors, equals 0 both when $N = 0$ and when $N = a/b$ (see Bøggild and Keiding, 1958). Consequently, initial numbers of individuals that approach a/b threaten a population with extinction.

For many plants and animals the initial number of zygotes in any generation exceeds by far the number of adults the environment can support—that is, the carrying capacity of the environment. In some manner the large number *must* be reduced to the smaller one. In gardening this is done by "thinning." Nearly every packet of seeds bought at the local garden center carries an admonition: "Thin young seedlings to one for every 2 [or 4, or 6] inches." I suspect that this admonition is of greater importance today than it was fifty years ago because seeds of many garden vegetables are now of hybrid origin. The uniformity of modern-day hybrid seedlings removes much of their former ability to thin themselves.

Foresters have in the past often relied on young trees to cull themselves. Individuals of certain species were known to establish clear-cut dominance relationships more easily than those of other species, which required hand thinning. Furthermore, if hand thinning were postponed too long, an entire planting might have to be destroyed because the roots of the crowded trees would already have become matted and stunted. In commercial stands that are tended as other farm crops, an annual harvesting of a portion of young trees takes the place of the earlier self-culling.

Phenotypic variation, then, provides the means by which untended populations of organisms—animals and plants—cull themselves so that the number of individuals present at any moment corresponds (at least approximately) to the number that the environment can support. This variation, of course, is phenotypic variation with respect to the ability to survive in the presence of competitors: some individuals survive while others die. The implications of this claim should be clearly understood:

- The ability to compete successfully with one's neighbors is an important component of individual fitness.
- Persistence requires that organisms be able to cull themselves so that their total numbers at any moment match the carrying capacity of the environment.

- Efficient culling (culling that leaves the victor essentially unharmed) requires phenotypic variation with respect to competitive ability.
- Phenotypic variation with respect to individual competitive ability results in a *phenotypic load* because the average (relative) competitive ability of a population must be lower than that of the optimal phenotype. It is this load, however, that enhances persistence and, on that account, *increases the fitness of the population*. This claim—that a load can increase fitness—is diametrically opposed to the view of genetic load espoused in the earlier chapters of this book. According to Haldane, Muller, and others, variation lowered fitness by an amount related to mutation rate (the source of genetic variation). Now, however, because of the implications of soft selection and the need for greatly reducing the large initial numbers of young zygotes, variation with respect to competitive ability is said to enhance a population's ability to persist through time, to enhance the population's fitness.

Phenotypic variation with respect to competitive ability need not be based entirely on genetic variation. Age differences result in phenotypic variation. A mature maple, elm, or sequoia produces thousands (even millions) of seeds each year. These seedlings pose no threat to the parental tree, which is firmly established as a feature of the environment.

News from disaster areas or famine-stricken regions of the world confirm for human populations that the ability to survive is not distributed equally among all age groups; the very young and the very old are those most often viewed as moribund on the evening television newscast. Dobzhansky (personal communication) during the late 1950s believed that natural selection for improved homeostasis would lead to a relatively uniform physical "fitness" extending from the prime of one's life through middle and old age; death would arrive as an abrupt, albeit natural, termination of the individual's life. To the extent that a uniform ability to withstand stress (extending well into advanced years) might lead to aged, sterile survivors following a prolonged stressful period, I suspect that natural selection would not promote uniformity in fitness, or at least not in that component of fitness representing ability to survive under competition. Among individuals of natural populations, this component would, I suspect, be greatest among advanced adolescents and young adults of both sexes. The presence of such favored individuals would be more nearly ensured in a population of overlapping generations than in one where reproduction is largely synchronized.

The other important source of nongenetic variation is chance, under which I include both microvariation of the environment and accidents that befall developing individuals. Among plants, one can speak of poor and good patches of soil—rocks and loam—upon which seeds fall by chance and subsequently develop. Of two neighboring plants, the larger will shade and eventually eliminate the smaller. In long-lived species, the arena within which pairwise competition occurs becomes steadily larger. If in a forest all leaves are removed from a sapling in order to demonstrate as a "classroom" exercise that such an event need not be fatal to a growing tree (it will put out new leaves the following year), the demonstration is valid only for the short term: the lack of one year's growth places the experimental sapling under a severe handicap vis-à-vis its nearby competitors and, in all likelihood, will lead to its eventual demise.

Leigh Van Valen (1975) has reconstructed the events leading to a

Table 8-1. Life-table statistics for the palm *Euterpe globosa*. (After Van Valen, 1975.)

Stage	Height (m)	Age (years)	No. per 5926 m² per year-class	Surviving proportion of initial cohort
Seed on tree	—	0	170,000	1
Seed on ground	—	0.1	46,000	2.7×10^{-1}
Germinated seed	—	1	2200	1.3×10^{-2}
End of seedling	0.5	9	205	1.2×10^{-3}
Immature tree	1.0	15	80	4.7×10^{-4}
	2.0	24	40	2.4×10^{-4}
	3.0	30	18	1.1×10^{-4}
	4.0	36	7	4×10^{-5}
	5.5	45	3	2×10^{-5}
Reproductive maturity	6.5	51	0.5	3×10^{-6}
	9.0	66	0.2	1×10^{-6}
In canopy	12.0	88	0.2	1×10^{-6}
	14.0	104	0.5	3×10^{-6}
	16.0	130	0.3	2×10^{-6}
Senescent	18.0	156	0.1	6×10^{-7}
	20.0	182	0	0

Self-culling and Persistence

stable number of adult palm trees (*Euterpe glabosa*) in a grove (Table 8-1 and Figure 8-1). In examining Van Valen's data, it is important to note that each item in the table and each point in the figure is independent of all others; that is, these data were not obtained by following a single cohort. Data of the latter sort produce artificially "smooth" curves because the number of survivors at any time must be smaller than (at least, no larger than) the number at the preceding stage.

The data on palm trees reveal the tremendous thinning that must take place in reducing an initial number of zygotes to a number of adult individuals (fewer than 2 trees per 5926 m²) that is compatible with the carrying capacity of the environment; fewer than 1 of every

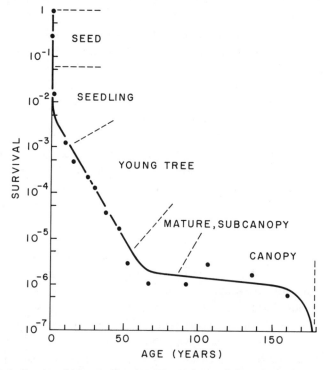

Figure 8-1. Survivorship curve for the palm *Euterpe globosa*, extending over seven orders of magnitude, illustrating the extent to which initial cohorts of this species are thinned before the stable population of mature plants (matching the carrying capacity of the environment) is attained. (After Van Valen, 1975, in Wallace, 1981, courtesy of Columbia University Press.)

100,000 zygotes survives. If genetic load is to be useful in this case as an ecological concept, one must postulate that this enormous elimination is independent of that which results from genetic variation.

A phenotypic load is needed to promote the efficient self-thinning of a population, a culling that reduces the number of individuals to correspond to the carrying capacity of the environment. The phenotypic load results from phenotypic variation with respect to fitness. As we have seen, this variation may stem from nongenetic agents: age and microenvironmental variation. However, genetic variation may also lead to phenotypic variation. To the extent that genetic variation contributes to the phenotypic variation that in turn promotes the self-thinning of populations, genetic variation enhances persistence, thus *in*creasing the fitness of populations. By definition, we may recall, genetic variation results in a genetic load. Therefore, to continue our reasoning, *a genetic load may enhance a population's fitness.*

The genetic load calculated by mathematical geneticists arises as a consequence of averaging the dissimilar Darwinian fitnesses that characterize the individual members of a population. The fitness of a population, under this view, is maximized when the Darwinian fitnesses of its individual members are at identical maxima. In the case of Van Valen's palms, however, such perfection would not save the 99,999 individuals of every 100,000 that must be culled in order to reduce the number of seeds and seedlings to the number of adults that the local environment can support. Indeed, the thinning would be made much more difficult, or even impossible, by such unvarying uniformity.

Under the genetic load concept, a single survivor among the palms would be chosen at random from every initial 100,000 seeds. Under frequency- and density-dependent selection—soft selection—the elimination of seeds and seedlings would begin with the genetically weak and environmentally unfortunate zygotes; each elimination would enhance the probability of survival for the remaining individuals. In the grove of palms studied by Van Valen, 1 individual of each 100,000 survived; under harsher conditions in another grove, self-thinning might leave only 1 individual of 200,000, 500,000, or even 1,000,000 seeds. The process of natural selection, however, would remain unchanged; selective elimination begins at the low end of the phenotypic fitness scale and proceeds upward (rank order) until the number of still-unculled individuals corresponds to the number the environment can support. At that time, the probability that each existing individual will become (if it hasn't already even as culling proceeds) a reproducing adult is very nearly 1.00.

Genetically based phenotypic variation with respect to survival under crowded, competitive conditions can be classified as follows:

1. Different genotypes give rise to different phenotypes; self-thinning, in this case, is accomplished by the selective favoring of certain genotypes—not necessarily the same genotypes under different conditions, at different localities, or during different years, however.
2. A given genotype might, through developmental plasticity or individual physiology (especially in plants), give rise to a phenotypically variable collection of progeny zygotes.

The first of these two categories can be conveniently subdivided as follows:

1a. The alleles responsible for the phenotypic variation may occupy essentially unlinked loci (*polygenes*) and, upon recombining each generation, may produce a broad unimodal distribution of individual fitnesses, of which some (with respect to a given environment or combination of environmental factors) are especially poor, some are especially good, but most are about average.
1b. Through the use of chromosomal inversions, translocations, or other devices that prevent the free recombination of genes at different loci, a population may be polymorphic. In this case the population may contain a collection of zygotes that tend to fall into discrete fitness classes (polymodal) whose averages are widely separated relative to the dispersal of individual fitnesses around each average. Nongenetic sources of phenotypic variation would tend to blur (but not necessarily obscure) the distinctions between the different classes, of course; even the healthiest of seedlings can be crushed underfoot or devoured by a hungry herbivore.

The unimodal distribution of individual fitnesses (particularly of the probabilities of survival under crowded conditions) is especially interesting with respect to the point under discussion: the relation of self-culling to population fitness. Several studies have suggested that individuals that are heterozygous at multiple loci have a selective superiority over less heterozygous individuals. Such evidence has been obtained by observing that within a cohort of young zygotes, the proportion of multiple-locus heterozygous individuals increases with time (see Singh and Zouros, 1978; Zouros et al., 1980). D. W. Garton et al. (1984) have shown for the clam *Mulina lateralis* that routine metabolic costs decrease with increased heterozygosity, thus leaving more of the individual's energy budget for reproductive effort. Such evidence, based on studies of natural populations, has been obtained for several marine organisms and for at least one tree.

The superior fitness of multiple-locus heterozygotes provides an exceptional opportunity for criticizing genetic load theory. It should be remembered (Chapter 5) that, according to load theory, a selective advantage of heterozygotes at many gene loci should overburden a population and lead to its extinction. Under soft selection, these heterozygotes are said to occupy the upper tail of the fitness distribution. As the too-numerous young zygotes of a population are reduced in number by natural selection, among those eliminated will be many of the ill-adapted homozygotes. Eventually, as the total number of surviving individuals is reduced, those remaining will consist largely of multiple-locus heterozygotes. There is no need for Mother Nature to count numbers of heterozygous loci in order to set a lower limit on adequate fitness; selective elimination starts at the lower end of the fitness distribution and stops only when the prevailing environment is adequate for supporting the remaining individuals.

The selective superiority of multiple heterozygotes possesses a second interesting and extremely important feature—entirely by luck, in my opinion. Mating among the reproducing, multiple-locus heterozygous individuals reconstitutes the starting population of zygotes: a highly variable population that is distributed around a single mode. The pattern of selection that was described in the previous paragraph, consequently, can be repeated again and again, generation after generation. The selective superiority of any one homozygous genotype would, in time, lead to a genetically uniform population; it would then lack that genetic component of the phenotypic load which I claim promotes self-thinning in natural populations. (See O'Brien et al.,

Self-culling and Persistence

1983, for an account of the virtual genetic identity of all living cheetahs and for a discussion of associated problems.) However, because multiple-locus heterozygotes seem to be the selectively favored individuals (for whatever reason; see Wallace 1975b, 1976; Wallace and Kass, 1974; Turelli and Ginsberg, 1983), this loss of genetic variation is avoided.

If my argument has been stated clearly, then the contrast between it and an account presented by Stebbins (1958) should be obvious: Stebbins claimed that long-lived trees that produce many seeds can afford the wastage of seeds that accompanies polymorphisms at many loci. The present argument, in contrast, says that the phenotypic variation in fitness to which genetic polymorphisms contribute enables the initially huge numbers of zygotes to be culled to a number (including *one*) of adult trees commensurate with the carrying capacity of the local environment. The question, consequently, is not one of *affording* what is an otherwise harmful characteristic, but of *enabling*. That among the many zygotes that are killed annually by the continued presence of a long-lived parent tree are some potentially highly fit individuals does not demonstrate (as some forest ecologists would believe) that natural selection is inoperative in populations of trees. The effectiveness of selection is manifest in the competition among seedlings and young saplings following the death of an old tree; only one individual will eventually succeed in occupying the newly created space in the forest (Figure 8-2). If the successful candidate tends to be a multiple heterozygote, then such genotypes can be said to possess high fitness.

Different population and community problems call for different strategies on the part of natural populations. The discussion in the past few pages has dealt primarily with *intra*specific competition—the culling of many zygotes to a number corresponding to the carrying capacity of the local environment. At times, however, the challenge to a population's continued existence comes not from a failure to thin its own numbers but from a need for its members to compete successfully with members of a different (usually closely related; often congeneric) species—that is, *inter*specific competition.

In the face of intense interspecific competition, a unimodal distribution of fitnesses of which only those in the upper tail are exceptionally high may not be a proper or appropriate strategy. Much better might

Figure 8-2. Two stages in gap phase replacement in a maple-basswood forest. The upper sketch shows a dense stand of maple (*Acer saccharum*) seedlings, about two to three feet in height, growing in a forest clearing. The lower sketch, illustrating a clearing about ten years older than the first, shows the still-numerous but reduced number of saplings that are competing to fill the gap. Seedlings of any age in the latter case are virtually absent. The different heights of the still-competing saplings suggest which one of them will be the eventual survivor. On the other hand, the removal of all but a few of the smaller saplings would not prevent one of the latter from becoming that survivor. (After Bray, 1956.)

be a reliance on a genetic mechanism that permits the production of relatively large numbers of uniform, highly fit individuals. The culling in this case is not largely of one's conspecifics but, on the contrary, of individuals belonging to a competing species. The chromosomal polymorphisms of many species of plants and animals—especially the inversion polymorphisms of many *Drosophila* species—appear to be devices that allow the population to take the greatest feasible advantage of hybrid vigor. Each generation of zygotes includes a sizable number of highly uniform, highly fit individuals that are heterozygous for chromosomes carrying different gene arrangements; the structural homozygotes tend to be less fit. If the heterozygotes were to compete primarily with one another, the uniformity of individuals would offer little, if any, basis for culling. However, in an area where a closely related species of the same genus exists (or any other species making nearly identical demands of the environment), the necessary culling is largely of one's competitors. I have discussed the contrasting patterns of structural homozygosity and heterozygosity of plants and animals in the light of the immobility of individual plants, and the differing community problems encountered at the centers and margins of species distributions (Wallace, 1983). Whether *Drosophila* species become structurally homozygous at the species border or retain most of their structural heterozygosity may depend upon the factors that have given rise to the border: Have suitable habitats vanished because of subtle changes in the environment? Or have otherwise suitable habitats "vanished" because they are occupied by a closely related *Drosophila* species whose (potentially overlapping) distribution abuts on that of the first species (as in Figures 6-9 and 6-10)?

In classifying the genetic basis for variation that might be useful in promoting self-culling, a uniform genotype capable of producing variable progeny was mentioned. If the one-genotype–one-phenotype pattern is regarded as revealing *proximate* genetic variation, the one-genotype–many-phenotypes pattern might be regarded as revealing *ultimate* genetic variation. In the latter case, the selectively favored phenotypic attribute of the successful genotype is said to be that it produces variable progeny.

Variable progeny produced by a single genotype? What sort of genotype would that be? These questions are best answered by an example. Everyone has heard of the proverbial sameness of peas: as

Table 8-2. The proportions of vestigial (aborted), small, and normal-sized garden peas at various positions in the pod (position 1 is at the peduncle end). (After Linck, 1961.)

Size	Position in pod						
	1	2	3	4	5	6	7
Vestigial	59	22	0	0	19	41	54
Small	41	59	41	12	31	44	46
Normal	0	19	59	88	50	16	0

alike as two peas in a pod. What, though, are the facts about peas in pods? The results of a study by A. J. Linck (1961) are shown in Table 8-2 and Figure 8-3. The peas in many pods of a given variety of garden peas were counted and classified as vestigial (aborted), small, and normal sized; they were also identified as to their site within the pod (site 1 was near the peduncle). The results are clear: peas in a pod are not alike. The proximal and distal sites tend to have small or aborted seeds, whereas the central sites tend to be occupied by normal (larger-sized) peas.

Accounts similar to the one mentioned above for peas could be cited for many plants. Kernels of corn on a cob are not uniform in size, nor are the kernels of different cobs of the same average size. Grains of wheat borne by secondary tillers are smaller than those borne by the main part of the plant. Figure 8-4 illustrates the fruit borne by a silver maple (*Acer saccharinum*); this tree is of a mutant variety that appears to be spreading within the species. Most, but not all, pairs of fruits on this tree have one member aborted; the remaining single fruit of each singlet is considerably larger on the average than either of the two fruits of a "normal" pair. J. L. Harper (1977) devotes many pages to a discussion of *somatic polymorphism* in plants; much of his discussion concerns somatic polymorphism and dormancy. Seeds and fruits of plants vary (*within* plants) with respect to size, color, hardness of seed coat, and control of germination time. Such variation seems to be more the rule than the exception.

Evidence that the observed variation among progeny of a given individual (a single genotype) has an ultimate genetic basis is indirect, yet rather compelling. Artificial selection can result in more uniform

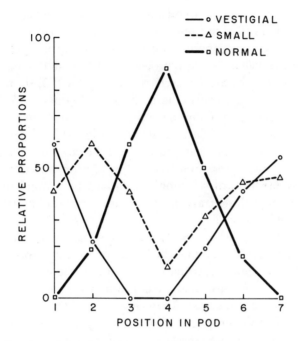

Figure 8-3. Peas in a pod are not all the same size. Solid line and circles, vestigial or aborted peas; dashed line and triangles, small peas; heavy line and squares, normal-sized peas. (After Linck, 1961, in Wallace, 1981, courtesy of Columbia University Press.)

Figure 8-4. The systematic abortion of one member of many paired fruits of a particular maple (*Acer saccharinum*) tree. The single remaining fruit is consistently larger than either of the two fruits of the "normal" pairs. Two-membered fruits that germinate without separating probably produce only one viable seedling (see Figure 6-5). Variation that occurs within individual plants as illustrated here is referred to as somatic polymorphism (see Harper, 1977). (From Wallace, 1982, courtesy of *Biological Journal of the Linnean Society*.)

peas in a pod, kernels of corn on a cob, and grains of wheat on a wheat plant; indeed, much of the time and energy expended by plant breeders is devoted to creating more uniform agricultural products—uniformity within as well as between progenies. Success in lessening within-progeny variation by artificial selection (see Table 8-3) shows that the initial variation has a genetic basis. Kenneth Mather (1953) demonstrated that even variation in the numbers of sternopleural bristles on the right and left sides of *Drosophila melanogaster* has a genetic basis; such variation (developmental noise?) can be either increased or decreased by artificial selection. Interestingly, after selection in either direction, relaxation of selection is followed by a rapid restoration of the average right-left asymmetry characterizing the original, unselected population of flies.

The reproductive behaviors of birds can be cited as a further illustration: Some birds (ducks and chickens, for example) produce large clutches of eggs before starting incubation; all the young of a clutch thus hatch simultaneously and are uniform in age and size. Synchrony of hatching, as ornithologists know, is enhanced by egg-to-egg verbal communication. Other species of birds lay eggs at several-day intervals and begin incubation with the first egg laid; the offspring in these cases are of different ages and sizes. *Cainism*, the killing (and often eating) of one sibling by a larger one, is common among the latter bird species; the different sizes of combating sibs makes an extended physical struggle both unnecessary and unlikely. That different species of birds exhibit such different reproductive behaviors demonstrates that neither pattern is *necessary* (that is, unavoidable). Presumably the differ-

Table 8-3. The relative proportions of hard seeds on individual plants (*Trifolium berytheum*) collected from wild populations or subjected to unintentional selection during eight years of domestication. Hardness of seed is related to speed of germination. (After Katznelson, 1976.)

	Proportion of hard seed									Total plants tested
	0.10	0.20	0.30	0.40	0.50	0.60	0.70	0.80	0.90	
Wild	0	0	0	0	0	1	1	4	17	23
Domesticated	7	6	6	4	4	6	8	7	2	50

ent patterns have arisen through natural selection; that is, the different patterns have *ultimate* genetic bases.

Persons may argue, especially in the case of plants, that rather than the plant's genotype, the anatomy of the growing plant, the distribution of nutrients among tillers, and the patterns of shading among leaves determine the observed variation among seeds and fruits. Such reasoning has prompted the use of the terms *proximate* and *ultimate* in referring to the role of genetic mechanisms in determining variation. An analogous problem arises in connection with the onset of the breeding season in many bird species. Physiological experiments reveal that changes in the number of daylight hours trigger many birds' reproductive behaviors; photoperiod, then, is the proximate cause of sexual activity. Photoperiod, however, merely provides a signal that, if properly read, synchronizes the birth of nestlings and the availability of food for feeding them. Different species utilize different signals for this purpose. Thus, the genetic constitutions of different species of birds, operating through different sorts of signals (of which photoperiod is only one), are *ultimately* responsible for coordinating the need to feed young nestlings with the availability of appropriate food.

In an effort to disentangle physiological and genetic mechanisms (that is, proximate and ultimate causes), I presented the argument that is outlined in Figure 8-5 and Table 8-4 (Wallace, 1982). Figure 8-5 illustrates several pairs of hands and feet. No one disputes that the development of an individual human being is under genetic control. It is no accident that our forelimbs terminate in hands and that our hind limbs terminate in feet. Nor are the patterns of fingers and toes arrived at by accident. Virtually every detail of limb formation is under genetic control; gene mutations that result in webbing between digits, short digits, fused digits, extra digits, total absence of hands or feet, and still other abnormalities are well known.

If the variation among digits on an extremity (fingers on hands, toes on feet), between extremities (in which fingers would be contrasted with toes), and between individuals were statistically analyzed, all variation would be accounted for by the dissimilarities of individual fingers and toes, and by the contrast between fingers as one group and toes as another. Virtually no variation would be ascribable to differences between individuals; this lack of interindividual variation re-

Fifty Years of Genetic Load

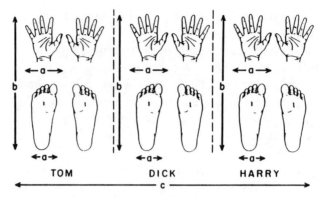

(a) VARIATION BETWEEN DIGITS WITHIN EXTREMITIES

(b) VARIATION BETWEEN EXTREMITIES WITHIN INDIVIDUALS

(c) VARIATION BETWEEN INDIVIDUALS

Figure 8-5. An illustration of hands and feet that emphasizes the variation between digits (fingers and toes) on the same extremity and between those on different extremities (fingers vs. toes). Because these differences are ascribed to the genetic program controlling the development of human beings, similar variations in plants (somatic polymorphisms) may also be ascribed to the (genetic) developmental programs of plants. (From Wallace, 1982, courtesy of *Biological Journal of the Linnean Society*.)

flects the overall similarity of the genetic programs that control human development.

Table 8-4 illustrates the results obtained when the argument outlined above was applied to nine sweet-pea plants collected from a population that had grown wild in pastures (first in Pennsylvania and, later, in New York) for thirty-five years. After drying, the peas of these plants were individually weighed; the data were recorded according to both pods and plants. The analysis of the data corresponds to the one described above for fingers, toes, hands, feet, and individuals. The result, as in the case of human beings, reflects a common genetic program that governs each plant's development. In Table 8-4 we see that considerable variation exists between peas in a pod, and that a comparable amount exists between pods on individual plants. Virtually no variation remains, however, in differentiating the peas borne by different plants. The suggestion, then, is that a common genetic

Table 8-4. Estimated variance components with respect to the weight of peas borne by nine "wild" sweet-pea plants. (Wallace, 1982.)

	Variance*		
	A	B	C
Source of variation			
Between peas; within pod	176.66	186.80	187.29
Between pods; within plant	155.26	134.73	137.15
Between plants	−3.37	7.98	0

*Estimated as follows: A, MIVQUE (0) estimation procedure; B, variance components obtained from ANOVA; C, maximum likelihood estimation procedure.

program governs each pea plant's development, and that this program generates the observed variation among progeny seeds.

An important item concerning the source of the phenotypic variation among the individual members of single progeny remains to be discussed; it concerns the possible need to invoke group selection as the only means by which a genotype capable of producing variable progeny can be favored. In the case of a proximate genetic basis for phenotypic variation, we might recall, the selective superiority of multilocus heterozygotes was sufficient to account for both the presence of selectively superior individuals in each generation and (fortuitously, it seems) a pattern of inheritance capable of regenerating the entire spectrum of variants, generation after generation, through the mating of surviving individuals.

A claim that selection might favor a genotype that, as one of its phenotypic properties, produces variable progeny, some of which must be selectively inferior to others, appears to contradict expectations based on Darwinian selection. Those expectations are that individuals that produce progeny of uniformly high fitness will displace those that produce progeny some members of whom are selectively inferior: whereas all of the progeny of the one plant might survive, for example, only some fraction (by stipulation) of the variable progeny of the other would survive. If that fraction were $\frac{1}{2}$, the relative Darwinian fitnesses of these two plants would be two to one.

Omitted from the above account of expectations, however, is the threat of overcrowding that occurs because of fluctuations in the carrying capacity of the environment. In the top diagram of Figure 8-6 are two plants, one of which produces uniform (black) seeds and the other variable (white) seeds. In this diagram the seeds of each plant are shown as having fallen at the base of their parent. Even in the best of times these seeds undergo severe culling; their numbers are too great to allow them all (or even more than a single individual) to survive. In addition, because of variations in the local environment's carrying capacity, even more severe self-culling may be required at times. The uniformity of the progeny on the left makes all culling difficult; the variation among seeds on the right eases this task. Although the composition of the assemblage of progeny individuals determines the likelihood that at least one will survive, selection in this case is none other than Darwinian selection: the parental genotype that generates surviving offspring is the genotype favored by natural selection.

The lower diagram of Figure 8-6 illustrates a case that is more difficult to analyze: the seeds (or fruit) of the two plants are intermingled in a seemingly random fashion. That seeds might be intermingled in a truly random fashion is, of course, unlikely; seeds, fruit, fallen leaves, and even pollen grains are much more numerous at and near the plant that produces them than they are elsewhere. Consequently, most mixing of seeds would result in a mere modification of the top diagram (whose evolutionary outcome conforms to Darwinian selection) rather than in the creation of an entirely new problem.

Even granting, however, that the seeds may be thoroughly, and randomly, intermixed in a restricted area, there is no need to abandon an individual (i.e., Darwinian) explanation for the superior fitness (as parents) of plants that produce variable offspring: the offspring that survive(s) during a period of stress may represent any extreme of a multidimensional range of variation. Seed size is an obvious variable to illustrate (Figures 8-3 and 8-6), but variation includes many additional traits. Harper (1977:69) describes the conjoined pairs of seed in *Xanthium* species. These seeds are dispersed together, but "at least 12 months normally separate the germination of the two seeds in each dispersal unit." The points made here were also made by M. Westoby (1981) in an article titled "How Diversified Seed Germination Behavior Is Selected." The need to invoke group selection is avoided by ascribing to the mother plant the selective events occurring among progeny individuals. I believe that one would seriously err in thinking

Figure 8-6. Selection favoring genotypes that result in somatic polymorphisms (represented here as seed size) need not be *group* selection but, on the contrary, is probably *Darwinian* selection. This is obviously true in the upper diagram, where the nonpolymorphic and polymorphic individuals are shown in isolation; the variable fruit of the one allows for self-culling, whereas the uniform fruit of the other does not. The situation illustrated by the upper diagram is altered only quantitatively by mixtures that still allow the bulk of a plant's fruit to be found directly beneath or in the immediate neighborhood of that plant. However, recalling that the somatic polymorphic variation occurs in many dimensions (seed color, dormancy, hardness, and size, for example), the individual most likely to escape destruction by overcrowding in the lower diagram, in which the fruits are thoroughly mixed, is still one of the variants produced by the somatically polymorphic plant. (From Wallace, 1982, courtesy of *Biological Journal of the Linnean Society*.)

that a plant that produces seeds of uniform color, which germinate simultaneously and from which seedlings exhibiting uniform growth arise—that such a plant would be favored by Darwinian selection (see Figure 8-7 for an example drawn from fish).

Indeed, the problem under discussion can be inverted: somatic polymorphism in plants is a common phenomenon that either has or does not have a genetic basis. Changes, even unintentional ones, in the nature of these polymorphisms that arise during domestication (see Table 8-3) demonstrate that they do indeed have a genetic basis. This variation (*ultimate* genetic variation in my terminology) may be favored by Darwinian selection as I have argued, or it may have arisen by some other pattern of selection—group selection, for example. *If somatic polymorphisms are common and if I have erred in my analysis, then group selection cannot be as unimportant in leading to evolutionary change as many persons believe.*

In summary, a case has been made in this chapter arguing that phenotypic variation with respect to competitive ability increases the fitness of a population by facilitating self-culling and, as a result,

Fifty Years of Genetic Load

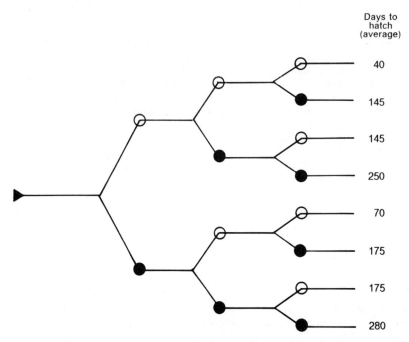

Figure 8-7. The six major developmental times of annual fishes that arise from the existence of three prehatching, nonobligate diapause stages. A certain proportion of all developing embryos escapes each diapause, the first of which lasts 30 days and the remaining two, 105 days each. Problems confronting these fish (and the plants illustrated in Figure 8-6) are multidimensional, thus precluding the adoption of a single best response to an inconstant environment. Fertilization is represented in the diagram by the solid triangle on the left; each diapause is represented by a solid circle; individuals escaping diapause pass through the open circles. (After Wourms, 1972.)

decreasing the probability that overcrowding will cause the population's demise. This argument stresses that persistence is a component of a population's fitness. It also implies that although persistence is a trait that can be verified only in the future, its present causes are not unfathomable.

Personal comments

The controversy involving the "balanced" and "classic" views of population structure is now of sufficient age to attract the attention of historians. Not that the issues have been settled; on the contrary, many are with us still. The world has moved on, however. New analytical procedures have opened up new avenues of research, and many young workers show no interest in either of the disputed views. Within a decade those with firsthand knowledge of the controversy, which was largely a dispute between Dobzhansky and Muller, will be gone; within another decade those who were merely acquainted with the disputants will be gone as well.

I recently read a manuscript written by John Beatty (see Beatty, 1987), who has undertaken to prepare a historical account of the Dobzhansky-Muller dispute. I do not envy him because, as a historian, he feels compelled (indeed, he *is* compelled) to determine *why* a scientist takes a particular position on an issue at a given time. I generally find it sufficiently difficult to interpret *what* a person means from what he or she has written; any attempt to understand the *why* of a particular manuscript would be beyond my ability.*

One of the contrasts between Dobzhansky and Muller that Beatty emphasizes is that Dobzhansky favored variation while Muller believed in uniformity. The reason for Muller's view is relatively clear: he believed in perfection—perfection for the individual and perfection for the population. The perfect population, in his view, was merely a collection of perfect individuals. Such perfection could be attained under natural selection, however, only if the environment were to remain constant and mutation were (eventually) to cease.

Dobzhansky's view is more difficult to identify. His views concern-

---

*An excellent example illustrating this point can be found in Crow (1987). On page 377 of that article, Muller is said to have argued strongly against the possibility that the normal allele is really a population of indistinguishable isoalleles. In Crow's words, "I think he always held firmly to this view." In Figure 4-1 of this text I attempt to illustrate Muller's view and, accordingly, refer to the *ideal* genotype as one that is homozygous for normal alleles at all gene loci. (Muller, as Crow underscored, placed great emphasis on the ubiquity of incomplete dominance; that is, nonnormal alleles supposedly lower the fitness of even their heterozygous carriers.) How, then, does the use of the term *ideal* constitute a caricature of Muller's views, as Crow suggests on page 366 of his article? The explanation lies in the *static* versus the *dynamic* interpretation of the term *ideal*. As an experimentalist, "ideal genotype" means for me "the best possible genotype that can be assembled today." Muller looked upon "ideal" as something that might eventually evolve, especially under conscious guidance (positive eugenics) in the case of human beings. Thus, although a historian of science might properly ask of me why I have used the word *ideal*, he or she may *not* ask me why I have chosen to caricature Muller's views, for I have not.

ing hybrid vigor and balanced polymorphisms obviously led him to accept variation as a necessary consequence of the selective advantage of heterozygous individuals. He could also argue that the selectively inferior homozygotes might better fit this or that portion of the environment, but, I suspect, this was merely making the best of a situation that was thrust upon him. Chromosomal mechanisms capable of generating populations consisting entirely of heterozygous or hybrid individuals are known, of course, but these were regarded by Dobzhansky and his contemporaries as "evolutionary dead ends." Dobzhansky would not have recommended, for example, that human beings adopt as a eugenic measure an *Oenothera*-type genetic system involving multiple chromosomal translocations. The diversity he favored among human beings was accepted, I think, because of the excitement that he personally experienced when confronted with a variety of personalities. His love of worldwide travel and different cultures, and especially his fascination with lush tropical rain forests, reflected his personal enjoyment of a varied environment. (Muller, in contrast, often attended scientific conferences carrying small electric heaters and thermometers with which he attempted to equalize the temperature in all corners of the hotel room in which he stayed.)

Despite his love of variation, I do not believe that the possible adaptive role of phenotypic variation in *reducing* the sizes of overcrowded populations through self-thinning occurred to Dobzhansky. His instincts would have been to stress the selective advantage residing with large populations, not those residing with populations commensurate in size with the carrying capacity of the environment. Numbers of individuals and biomass were for him legitimate measures of population fitness. And, I must admit, endangered species are identified even today as species whose numbers are decreasing—not as those whose numbers are growing.

9

SUMMARIZING REMARKS

In the preceding chapters I have attempted to trace my involvement with and my changing views of the genetic load concept. I have credited Haldane (1937) with originating the notion that in gaining genetic variation, a population pays a price. Furthermore, the price is a simple function (multiple) of the rate at which one allele mutates to another.

Whereas the early, naturalist-type population geneticists (under this term I would include Dobzhansky, Dubinin, Tchetverikov, and the Timofeeff-Ressovskys—all Russians) spoke for years of the genetic variability that lay concealed behind a façade of phenotypic uniformity, it was H. J. Muller and his expression "our load of mutations" that stirred concern for the genetic structure of human populations—especially with the advent of the atomic (now nuclear) age.

From the outset, the genetic load of a population was seen as an undesirable burden, much like the one borne by Pilgrim during his search for salvation in Bunyan's *Pilgrim's Progress*. As with any new concept, genetic load was pulled at, poked, teased, tested, and manipulated by a host of geneticists. Loads were classified. Relative contributions to loads were calculated. The sizes of bearable loads were discussed under the term *load space*. A colleague who learned during a private conversation at the United Nations during the 1950s that the M-N blood group polymorphism in human beings is apparently maintained by the selective superiority of MN heterozygotes exclaimed in obvious relief: "Good! Now there is no more load space; there can be no more balanced polymorphisms in man!"

Unfortunately for those who thought that the relationship between mutation and the average fitness of populations should be a simple one, data collected from populations of *Drosophila* whose mutation rates were altered by exposure to mutagenic radiation did not confirm the algebraic calculations. This did not prove that the calculations were wrong, as Sewall Wright once carefully explained to a small audience at the Argonne National Laboratory; it merely showed that there are genetic effects that were not taken into sufficient account in the earlier calculations. For genes that behave as the calculations assume that they behave, the calculations are correct.

Matters became somewhat worse for mutational load adherents when it was found that randomly induced mutations, on the average, improved the egg-to-adult viability (and speeded up the rate of development) of otherwise homozygous individuals. This point is, as it might well be considering its importance to load theory, still a matter of debate. No test of this sort, it should be noted, has revealed a significant decrease (relative to simultaneous controls) in the viability of homozygotes carrying irradiated chromosomes. (I am, of course, discussing here the average effects of low levels of radiation; chromosome breaks are known to have an appreciable dominant and deleterious effect on the viability of heterozygous carriers; see Paget, 1954; Vann, 1966.) Three large tests have yielded significantly positive effects on viability. The remaining studies have shown both nonsignificant increases and decreases. Among these many ambivalent experiments are some whose methodologies are open to criticism (see Wallace, 1981:372).

In leaving the "formal" genetics of genetic load, we enter the realm of ecology and what is now called population biology. Here one can say that a population that maintains its size over a series of generations—neither growing nor shrinking appreciably—has an average fitness of 1.00: each mother (ignoring males) leaves one adult, breeding daughter as her replacement, on the average. If the value 1.00 is assigned to $\overline{W}$ rather than to the genotype with maximum or optimum fitness, the *relative* Darwinian fitnesses of different genotypes remain unchanged—only the numerical values assigned to them are altered.

The ratio of a population's size in one generation to that of the previous generation (best restricted to numbers of females in each generation) can be regarded as a measure of the population's fitness (see Penrose, 1949). Because it can be calculated only after the fact, and because it does not predict future events, this measure of fitness is of little practical use. The item that is lacking—the essential component of population fitness that is missing—is *persistence*: the probability that the population *will still exist* in the next generation.

Genetic load, as calculated by Haldane, Muller, and others, has little to do with fitness measured as the ratio of daughters to mothers (a ratio that equals 1.00 if the population maintains a constant size). Populations can grow in size while their genetic loads increase, and they can dwindle while their genetic loads decrease. Any selection that improves the fitness of heterozygous individuals within a balanced polymorphism, for example, increases the population's genetic load. This claim includes all cases in which there has been an evolution of dominance.

When a population is examined with respect to the replacement of one generation by the next, one realizes (as did Charles Darwin and A. R. Wallace) that for most species a tremendous number of individuals—offspring and former parents—are destroyed before reproducing or fail to reproduce if they do survive. These discarded individuals constitute the *ecological load* described by Turner and Williamson (1968). Individuals are destroyed by a variety of means, including starvation, predation, and accidents. They are also destroyed for genetic reasons; but genetic deaths and the other deaths are not independent. Individuals who suffer from a genetically caused weakness are more likely to starve or be devoured by predators than are others. *Genetic load is but one component of the total ecological load.* The specific assignment of deaths either to the environment or to genetic causes is largely arbitrary. Deaths caused by malaria can be assigned to the environment in some human populations and to genetic load in others, depending upon the presence or absence of the sickle-cell (Hb^s) allele. A similar arbitrariness exists in assigning deaths to environmental causes—let's say, to starvation or predation. A weak, starving individual is easy prey for a predator: How, then, should its death be classified?

A satisfactory study of a population's fitness cannot be made by merely calculating ratios between successive generations of the past. A somewhat improved study can be made, however, by manipulating (perturbing) the environment (increasing or decreasing the nutritive value of the growth medium; changing the nature of available nutrients; introducing parasites or competing species; altering the temperature, humidity, or ambient light; or changing other, nearly endless, factors) and noting the effect of these changes on the numbers of individuals in successive generations (including those that precede and follow the time of change). In a sense, such manipulations are made in an effort to determine the conditions under which the population can survive and those under which it becomes extinct: they are studies that attempt to distinguish between persistence and extinction.

The inclusion of persistence in the concept of population fitness drastically alters one's view of genetic load. Overcrowding is a constant threat to a population's continued existence. Surplus individuals *must* be removed (thinned or culled) either by hand in experimental populations and home gardens or by interactions among the competing individuals in natural populations. Failure to thin an overcrowded population can lead to the destruction or sterility of all individuals—to the population's demise.

Self-culling that is based on interactions between maturing individuals of a natural population proceeds most efficiently if these individuals vary with respect to their ability to compete. A clear-cut difference between two competing individuals leads to a prompt decision and to the infliction of the least harm on the victor. Thus, phenotypic variation with respect to competitive ability enhances the probability that a population will continue its existence by avoiding the grave danger of overcrowding. (The converse of this view leads to the designation of weight classes for many amateur and professional sports; the spectator's interest is enhanced by the prolonged struggle and uncertain outcome of contests between closed matched competitors.)

Phenotypic variation with respect to competitive ability can arise from nongenetic causes such as age and microenvironmental variation. It can also have genetic bases. The most obvious of the latter can be called *proximate* genetic variation; paraphrasing a well-known expression, one might say, "one genotype, one phenotype," although the reverse is obviously not true. Genetically caused phenotypic variation of the proximate sort arises because of the direct effect of the genotype on the phenotype. Here, then, a population's fitness is *increased* by the presence of a genetic load.

Summarizing Remarks

Some phenotypic variation can have an *ultimate* genetic basis; particular genetic constitutions may lead to the production of individuals whose progeny are preset to differ one from the other. Examples can be cited from birds that lay eggs at intervals under constant incubation (a practice that leads to offspring of different ages and sizes), plants that bear seeds of different sorts on different tillers, plants that bear two or more kinds of flowers (violets, for example), and plants that bear seeds of different sizes or seeds that germinate at different times. The varying developmental times of annual fishes (Figure 8-7) also illustrates this type of phenotypic variation. The genetic underpinnings of such patterns of reproduction can be inferred from interspecific differences in reproductive behaviors (ducks and chickens, in contrast to many birds of prey, incubate eggs only after the entire clutch has been laid) or by altering the prevailing (natural) pattern by artificial selection.

The suggestion that phenotypic variation with respect to competitive ability enhances persistence, and therefore population fitness, has a corollary: this variation should be clearly visible. Size is an especially important component of a higher animal's competitive ability, particularly in physical "combat." Differences in size, however, are not clearly reflected in the more easily observed linear dimensions of an individual. A 10% difference in size (regarding the individual as a sphere) amounts to only a 3% difference in linear dimensions. Allometric increases in the size of certain features of an organism during development, by increasing the variation in the sizes of these features, might facilitate self-culling; in the absence of such clarifying signals, a failure to recognize differences in size (mass) could lead to conflicts that seriously damage victors as well as destroy losers (Wallace, 1987b).

Personal comments

Much that I have written about self-culling must be ascribed to a conversation I had with two Canadian foresters, whose names I cannot recall, during the 1955 Cold Spring Harbor Symposium. These persons explained that pine forests consisting solely of F_1 hybrids obtained by interpopulation pollination did not produce the best lumber despite the heterosis exhibited by these hybrid individuals. When a cleared area in Canada was to be reforested, bales of seedlings were dropped from low-flying planes. These seedlings were planted by persons who marched side by side in a long line stretching across the barren area that had recently been harvested. Each bale of seedlings contained some that had been obtained by inbreeding, others that had been obtained by random crosses within local populations, and still others (a calculated percentage) that had been obtained by crossing individuals from isolated geographic localities. The latter were the most vigorous individuals—the interpopulation F_1 hybrids.

In the colorful language of foresters, weak seedlings, as they grow, "train" their competitors to grow taller. Each year some of the weaker trees die, but their presence until that moment has aided in the training of those that outlive them. Finally, at harvest time, the remaining trees for the most part are the superior F_1 interpopulation hybrids. However, these hybrids are not stunted as they would be if they had been overcrowded as seedlings, nor are they as bushy with lateral branches as they would be if only hybrid seedlings had been planted at a considerable distance from one another in anticipation of the space required by an adult tree. This account, which lay forgotten for many years, was my first introduction to the notion and consequences of self-culling.

Lest anyone has misunderstood, the suggested role of genetic load, as a contributor to phenotypic variation and, hence, an enhancer of persistence, applies to natural populations. Population geneticists discussing human populations have traditionally attempted to estimate the effect of this or that policy (standards controlling the exposure of workers to radiation or mutagenic chemicals) on the average fitness of a population, on $\overline{W}$. In retrospect, I believe this is a useless exercise. Emphasis should be placed, in my opinion, on the individual human beings, present or future, who either are suffering or are going to suffer as a result of radiation exposure. The average of relative, Darwinian fitnesses is not a suitable basis upon which to base decisions regarding the fates of human beings.

10

STILL TO COME...

In the previous nine chapters I have reviewed my association and eventual disenchantment with genetic load theory. The account I have given is undoubtedly much neater than it was in real life. For events that have occurred over an extended time—decades—"there" is always connected to "here" by a path that, when viewed in retrospect, appears simple. False starts, missteps, and entrapping cul-de-sacs are all but invisible from this retrospective point of view.

In the pages remaining I shall attempt to look not back, but forward. Where is population genetics going? How will the union of ecology and genetics into a population biology be effected? What will eventually constitute an adequate resolution to some present-day differences of opinion, especially with respect to the neutralist-selectionist controversy? I shall not attempt to provide sweeping global answers to these questions; such an attempt on my part would be completely out of character. Rather, I shall look at the present situation as an experimentalist, as one primarily interested in questions that might be answered promptly by properly designed studies. At the outset, I must admit that this entire exercise, this entire final chapter, may take place within a cul-de-sac that leads nowhere.

Where is population genetics going? Without attempting to speak for everyone who considers himself or herself to be a population geneticist, I can report that the road that I trod is now approaching its terminus. Genetic variation was—and is—the stuff of population genetics. At first, only two alleles were recognized at each of a few genetic loci in well-studied species. Later, multiple allelism was discovered; many loci in *D. melanogaster* proved to be "occupied" by numerous alleles. The trend throughout the history of genetics has been for the discovery of more and more alleles; by the 1940s the A-B-O and Rh blood-group systems had been refined by the recognition of A_1 and A_2 as well as the eight combinations of Fisher's *CDE-cde* complex.

The greater the number of alleles and gene loci available for study, the finer the classification of any population according to the genotypes of its members. Molecular techniques, both allozyme and DNA analyses, have transformed all organisms into potential genetic material in the sense that deer, moles, kangaroos, and sharks no longer need to be brought into the laboratory for controlled matings and pedigree analysis. DNA technology now allows for the unambiguous identification of individuals (and, consequently, of their parents and sibs).

Thus, I now foresee extended field studies, probably of small mammals such as voles and field mice but perhaps of small birds (or even fish) as well, in which the demography and genetics of populations will be followed in considerable detail for many years. Such studies may require the establishment of research parks where the populations can remain relatively unmolested for extended periods of time. They may also require a succession of committed investigators, a stipulation that will irk search committees at many universities. Nevertheless, many aspects concerning the ecology of populations and the role of natural selection in modifying these populations will remain mysteries until the relationship of all individuals, the movement of these individuals within the local area, and the genotypes of these individuals with respect to numerous gene loci are known with virtual certainty.

Biologists freely admit that the material with which they work in the course of their studies is unlike that of any other science. Almost without exception, living organisms differ from one another—not only those individuals that carry the same species designation but also those (sibs and cousins) that share family relationships. Worse! Current knowledge concerning differential gene action forces biologists to admit that different portions of one individual are (genetically) effectively different; this realization will be more readily accepted by botanists, I expect, than by zoologists.

Still to Come...

The term *genotype*, which has been usefully contrasted with *phenotype* by geneticists for decades, must now be used with greater circumspection (Figure 10-1) because the control of gene action may lie in the environment outside the individual and may terminate at appropriate gene loci in only a portion of an individual's cells.

Genotype in its conventional use refers to the sum of all genetic information that accrues to the zygote through the union of an egg and a sperm. Knowledge of an individual's genotype together with information on linkage relationships (including coupling and repulsion) allows one to predict the genic content of the gametes that individual will produce. The genetic information that accrues to an individual at fertilization is not necessarily identical with that which is used at various times during development in various cells, tissues, and organs, or that which may be activated in response to various environmental signals. The open arrows in Figure 10-1 indicate that the environment "enters" the individual in the form of gene-regulating signals, and, conversely, that the individual "enters" the environment by habitat-changing behaviors. In speaking of gene-environment interactions without further qualification, one tends to overlook that even a single individual may inhabit two (or more) environments. Rosie the Riveter of World War II fame provides an example (Figure 10-2). As a consequence of her job at an aircraft factory, Rosie's right arm (her riveting arm) became enlarged and muscular while her left one retained its slender appearance. Lobsters provide a corresponding example (Figure 10-3): one claw, either the right or the left, enlarges into a crushing claw; the other remains slender while growing long and tapered. If kept in a tank free of pebbles and other debris, both claws of a lobster remain slender. If provided with even a single pebble, the first claw (left or right) with which the animal manipulates the pebble will, through repeated manipulations and continuous exercise, develop into the massive, crushing one (Govind and Pearce, 1986; Govind, 1989).

Syracuse University was host to a symposium on population biology on June 7–9, 1967. In his introduction to the proceedings of those meetings, Lewontin (1968:4) confessed that "geneticists, embryologists, and ecologists remained untransmutated. But," he continued, "they did talk to each other... and a mutualism has grown, fostered in part by our symposium."

Despite this start, population biology has remained a science with

Fifty Years of Genetic Load

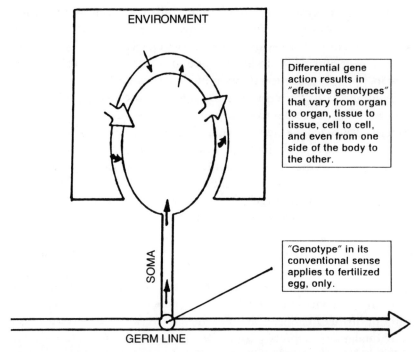

Figure 10-1. The development of the soma of a higher animal, well adapted to its environment, from a fertilized egg that lies within (and subsequently gives rise to the continuation of) the germ line. The large open arrows indicate that the environment "enters" the individual by means of signals that trigger gene control systems, and that the individual "enters" the environment by changing it in appropriate ways. The heavy solid arrows indicate inappropriate (nonadaptive) interactions between the individual and its environment, while the light arrows represent trivial interactions. The text suggests that because of differential, environmentally induced gene action within the cells and tissues of an individual, the conventional terms *genotype* and *genotype-environment interaction* fail to convey either the subtlety or the complexity of the actual events.

strong cleavage planes—sufficiently strong to clearly identify the population genetics and ecology sections of most textbooks. Recently, I wrote (Wallace, 1989a): "In retrospect, I feel that genetic load theory has been a severe hindrance to the development of a population biology that encompasses both population genetics and ecology."

How can the union of these two branches of biology be effected? That I shall not attempt to meld the two is the reader's good fortune; I would be unable to do so if I were to try. That an attempt has recently

Figure 10-2. A World War II poster depicting Rosie the Riveter, one of many women employed in the wartime aircraft industry. The "environment" that confronted this worker was complex; manual labor induced gene actions within her right arm that were not called for (at least to the same extent) in her left arm.

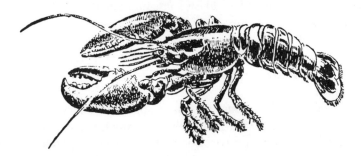

Figure 10-3. The lobster, *Homarus americanus*. Clearly visible in this sketch are the dissimilar claws, one of which is a powerful crushing tool while the other is better built for probing and tearing. The asymmetry (which eventually involves the central nervous system as well) develops after an animal starts using one claw (either left or right) to manipulate small pebbles; that claw becomes the large one. If raised in a tank without any detritus, a lobster will possess only small, tearing claws (Govind and Pearce, 1986; Govind, 1989). (Illustration based on sketch in *The Animal Kingdom* [Greystone Press, 1954].)

been made by someone more able than I is my good fortune. I refer to Adam Łomnicki's (1988) *Population Ecology of Individuals.** Both of us place considerable stress on individual variation, both genetic and nongenetic. I do not imply that Łomnicki is the only ecologist who stresses such variation, for, as he says (p. 66): "The adaptive significance of variation can also be explained by the concepts of the fitness set (Levins 1968), the evolutionarily stable mixed strategy (Maynard Smith 1982), or the optimal decision theory under unpredictable conditions (Cooper and Kaplan 1982)." The main reason I cite Łomnicki's book is the relative ease with which I follow much of his reasoning.

I feel no obligation to develop Łomnicki's arguments in this closing chapter; rather, I shall outline his views so that they might be compared with the ones expressed in the present text. He affirms, for example, that progress in ecology requires taking into account individual differences other than age and sex because properties of ecological systems must be derived from the properties of their elements (p. ix). Genetic variation may be one of the reasons for the existence of individual differences, but, he adds, it is not necessary for individual variation.

Łomnicki speaks, as I do, of the persistence of populations. If, for example, all members are identical, all will die simultaneously (see Table 7-1). In contrast, when large and well-pronounced differences among population members exist, some individuals will die promptly, leaving space and resources to others who survive and reproduce (p. 14). Population stability and persistence, he continues (p. 34), require not only unequal resource partitioning but also the inaccessibility to those of lower rank of resources controlled by higher ranking individuals.

Parallel to my own concept of unit spaces (Wallace, 1981:15ff.), which may be empty or within which only a single individual (the most competitive one) can survive, Łomnicki describes a patchy environment such that each patch is a place in which an individual organism

---

*Reviewers of Łomnicki's text (Mertz, 1988; Golenberg, 1989) have taken issue with much of his account of population biology. Flaws in his mathematical formulations are largely at fault. Even I sensed that in his final chapter Łomnicki appears to have neglected the individual variation that was an essential ingredient of arguments presented in early chapters. Nonetheless, many of Łomnicki's views are insightful, and his presentation is lucid. A widely accepted union of population genetics and ecology, however, is still to be effected.

can survive and leave progeny (p. 15). And where I speak of rank-order selection (Chapter 6), Łomnicki (p. 21) speaks of ranking individuals from the first, with the highest intake of resources, to the last, with the smallest intake. Each of us recognizes that the presence of individuals that affect the food intake of others is necessary for the skewed size distribution of individuals, and for increasing variance in size (p. 56). My hackneyed descriptive expression for this situation is: the rich get richer while the poor get poorer.

In concluding this account, I can cite (with obvious satisfaction) two more of Łomnicki's statements. The first (p. 34) merely confirms that age (one of my nongenetic sources of variation) can be an important source of intrapopulation inequality. The second statement corresponds well to points I stressed in Chapter 8. Individual differences, according to Łomnicki (pp. 58–59), are triggered (and enlarged upon) by differences arising during an early stage of life, reflect an ability to produce variable progeny, or arise as the result of hereditary differences in germination time or rate of early development.

Łomnicki emphasizes that the view he propounds renders Wynne-Edwards's suggestion (1962, 1987) of group selection as a means for regulating population size unnecessary. I have also suggested (Wallace, 1987c) that rank-order (soft) selection resulting from intrapopulation variation in competitive ability leads to population regulation without the need to postulate the *functions* and *genetic programs* that are otherwise essential features of group selection.

Considerable attention has been devoted in recent decades to what is known as the neutralist-selectionist controversy. Controversies generally serve but one useful purpose: they stimulate research. They tend to outlast that purpose, however, and drag on, perpetuated largely by the inability of the initial protagonists to agree in every detail and by the habit of those who are only peripherally involved to use the term *controversy* excessively.

The neutral theory (see Lewontin, 1974) does not claim that nearly all mutations are neutral; rather, it claims that many mutations are subject to natural selection but, being almost exclusively deleterious, are eliminated from populations. The theory does claim, however, that the bulk of *existing* genetic variation in populations (allozyme variation, for example) is neutral with respect to fitness; variation that is not neutral falls within the domain of genetic load theory. Only that portion of all genetic variation which has not yet been cleansed from the population can be observed, and that portion consists almost entirely of neutral variation.

Neutral alleles are not necessarily variants of genes with trivial effects; on the contrary, the physical absence (chromosomal deficiency or deletion) of a locus that is otherwise represented by numerous allozyme variants may be lethal. The neutrality of these existing variants can be formally described by saying that (in theory) any allele at that locus could be substituted for any other in any cell of any individual in any habitat without affecting that individual's subsequent development or eventual fitness. This description of neutrality is generally acknowledged to be overly stringent (as well as unverifiable). Consequently, neutrality (because it is a comparative concept) is assigned to two or more alleles if they *seem* to be neutral: they are neutral if their average fitnesses are equal (in large populations) or nearly so (in small populations). They are neutral if their fitnesses fluctuate with frequent reversals either through time, within patchy environments, within different background genotypes, or any combination of these three factors.

Two points that may have escaped the attention of many persons who are only casually interested in neutralism should be made. The first is of a philosophical nature concerning statistical analyses; it is illustrated in Figure 10-4. Virginia Tech lies at the south end of a residential street named Toms Creek Road; apartments that were erected beyond the city limits for (economically beneficial) tax purposes are situated at the north end. During early morning hours, the bulk of the traffic flows toward the university; in the evening, it flows homeward toward the apartments. At midday, traffic is considerably lighter than it is either in the morning or in the evening.

Figure 10-4 lists some (hypothetical, although realistic) data concerning numbers of cars and the directions in which they move. The highly significant differences in numbers of cars traveling north or south during the morning and evening hours allows us to construct explanations for these data: in the morning, students and others are going to the university for class or for work; in the evening, these same persons are going home.

The midday data are different: the numbers of cars going in the two directions do not differ significantly. Therefore, by convention we say

Still to Come . . .

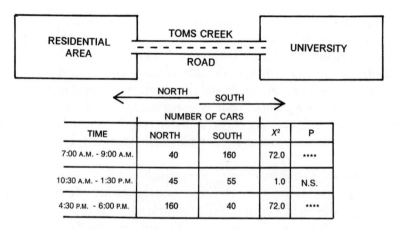

Figure 10-4. Statistically nonsignificant differences do not necessarily mean random or chance differences. The highly significant differences in the flow of morning and evening traffic lead to ad hoc interpretations: persons go to the university either to work or to class in the morning, and return home in the evening. The nonsignificant difference in traffic flow at midday does not mean that individual drivers are traveling at random; rather, it means that their reasons for being on the road take them north or south in comparable numbers. In brief, an inability to distinguish events from those expected by chance alone is no proof that those events are chance events.

that their movement can be accounted for by chance alone. This conclusion might lead some to conclude that midday drivers are guided at each intersection by the flip of an unbiased coin; hence the apparently random movement of noontime automobiles on Toms Creek Road. That conclusion, of course, would be absurd; the driver of each car at any time of day has a reason for being on the road and is intent upon reaching his or her destination. Randomness (statistical *non*-significance) does not imply absence of reason; on the contrary, it implies a multiplicity of reasons that in toto generate results that appear to be random. This statement can be paraphrased: randomness does not imply an absence of selection (i.e., neutrality in a strict sense); on the contrary, it may imply a multiplicity of selections that in toto generate effects that appear to be random. This conclusion can be expanded in making the second point.

Darwinian fitness of a single individual has little bearing on evolutionary change; that is a statement I made early in this book (p. 9). I made an analogous statement regarding dispersal (Wallace, 1981:35):

Fifty Years of Genetic Load

"A fly—one fly—can go anywhere by almost any means. . . . Rare accidents cannot be studied, but common ones can be. Although one cannot predict in advance the eventual movements or final resting place of an individual fly (or individual human being), one can describe the distribution of movements and the ultimate resting places of many flies." On a much grander scale we have the laws that relate volume, temperature, and pressure of gases; underlying the statistical nature of these laws are numerous molecules moving in seemingly random fashion.

Numerous molecules moving in seemingly random fashion. . . . But each molecule, during the interval of time separating two consecutive collisions, moves in a deterministic fashion, with a constant speed and a constant direction. The power of the gas laws and the statistical mechanics of which they are a part lies in the multiplicity of molecules, their velocities, and their directions; from these multiplicities comes the equation $V \sim T/P$.

In like manner, numerous flies (or mites, or human beings) moving in many directions at a variety of speeds lead to patterns that can be described in general terms, one of which may be that the logarithm of the number of individuals observed at a given distance from a point of origin decreases linearly with the square root of that distance. As was the case with molecules, descriptions of dispersal patterns are not concerned with individuals, or with the reasons that underlie the sequential movements of individuals.

The neutralist theory superimposes upon evolutionary change mathematical rules that are otherwise true for events occurring purely by chance and that would be true, as well, if the causes of nonrandom events were themselves numerous and varied. Here, however, one may be forgiven for pausing and urging caution in arriving at seemingly obvious conclusions: we might recall here that the midday drivers in Blacksburg, Va., are not mindless robots.

We might also recall that under Mendelian inheritance, and given that mutation and recombination occur, no two individuals of a population are likely to be identical. Variation between individuals is augmented by the environmental variation that these individuals encounter. The total variance among individuals equals the sum of their genetic and environmental variances, plus interaction effects.

Again, we might also recall both from previous chapters of the

Still to Come . . .

present text and from what has been cited from Łomnicki's (1988) *Population Ecology of Individuals* that the ecological properties of populations arise from individual differences. These differences ease the task of culling an enormous number of early zygotes to a number of adults commensurate with the carrying capacity of the environment. Furthermore, this culling, by preventing pathological overcrowding, leads to the survival of individuals that are capable of reproducing, thus perpetuating the population.

One more excuse for pausing and urging caution regarding the obvious conclusions based on neutralist theory can be cited. Gas laws deal with individual molecules that move about but remain individual molecules. Rules governing dispersal of organisms also describe the movements of individuals. Evolutionary "laws" deal with individuals any one (gravid female) of which can be the source of an exponentially expanding number of descendants. Seemingly trivial differences among individuals can, when events occur in proper sequence, become established differences between populations, species, or higher categories of individuals.

A discussion of the reservations outlined in the preceding paragraphs can begin with Figure 6-4, in which the stabilization of a population's size is related to the average number of adult daughters left per mother; the population attains a stable size when the ratio of daughters to mothers (D/M) equals 1.00. In populations composed of identical individuals (*AA* or *aa*) the equilibrium size can be represented as a *point* on the "base line" where the daughter-to-mother ratio equals 1.00. Polymorphic populations (those containing both *A* and *a* alleles), if the trajectories of the three genotypes differ, convert the *point* at which population size stabilizes into a vertical *line* through which the *AA*, *Aa*, and *aa* trajectories pass at different levels.

The three trajectories of *AA*, *Aa*, and *aa* individuals shown in Figure 6-4 are averages. Under different environmental (physical or biotic) situations, these trajectories may change. Hence, the vertical line representing stable population size in Figure 6-4 becomes a vertical *plane* as shown at the right in Figure 10-5. Only the near face of the three-dimensional diagram in Figure 10-5 corresponds to the average trajectories depicted in Figure 6-4. The trajectories that are characteristic of different situations ($S_1, S_2, \ldots$) differ in the heights at which they pass through the equilibrium plane. The average fitnesses projected on S_{avg}

Fifty Years of Genetic Load

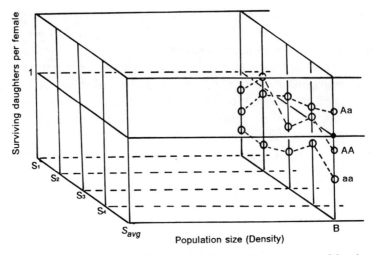

Figure 10-5. A population at equilibrium (B) with respect to size (ratio of daughters to mothers [D/M] equals 1.00) may possess individuals (*AA, Aa,* and *aa*, for example) for whom D/M ratios are not equal to 1.00 either in individual situations ($S_1, S_2, \ldots$) representing different environments and background genotypes or on average (S_{avg}). The solid circle that signifies an average of one daughter per mother lies in a plane (the *population arena*) within which the evolutionarily significant norms of reaction lie. (Compare with Figure 6-4.) (From Wallace, 1989a, courtesy of *Quarterly Review of Biology*.)

are obtained by averaging through all situations, each weighted by its relative importance.

The plane at the right in Figure 10-5 has been removed and rotated to present a face-on view in Figure 10-6. Population size is represented on the unseen axis, which is now perpendicular to the page of the text. The trajectories for the different genotypes (*AA, Aa,* and *aa*) pass into the plane of the figure at points shown for each situation ($S_1, S_2, \ldots$); the plane is the equilibrium plane at which the average number of daughters produced per mother equals 1.00 (indicated by the solid circle within S_{avg}). The complex curves connecting the plane-trajectory intersections for each genotype form the evolutionarily significant *norms of reaction* of these genotypes (see Schmalhausen, 1949; Lewontin, 1974; Gupta and Lewontin, 1982). The plane depicted in Figure 10-6 has been called the *population arena* (Wallace, 1989a). The qualifying expression "evolutionarily significant" is needed to differentiate norms of reaction that fall within the population arena from those that may be

Still to Come . . .

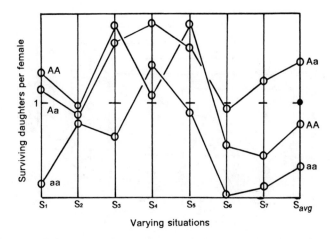

Figure 10-6. Full-face view of the population arena. The solid circle lying on the rightmost vertical line (S_{avg}) indicates that the population is at equilibrium with respect to size (D/M = 1.00). The different positions of *Aa*, *AA*, and *aa* individuals at S_{avg} reveal that, on average, mothers of these genotypes do not leave the same average numbers of daughters. The varying positions of open circles on lines representing different situations ($S_1, S_2, \ldots$) reveal that the relative fitnesses (as measured by daughters per mother) of the three genotypes vary from situation to situation. This variation leads to the evolutionarily significant norms of reaction for these genotypes. Fitnesses represented at S_{avg} are weighted averages of the fitnesses that are exhibited under various situations. (From Wallace, 1989a, courtesy of *Quarterly Review of Biology*.)

obtained under experimental conditions other than those characterizing an equilibrium population.

Population arenas can be arranged in time (Figure 10-7). The right border of each population arena (S_{avg}) falls within a plane that I call the *time arena*. This plane (or arena) reveals only the average number of daughters per mother (a measure of average fitness) for each type of mother. Studies conducted in this plane (unlike the norms of reaction that fall within population arenas) lack the depth of perception that allows an examination of relative fitnesses under different situations (some of which disappear with time, while new ones arise).

The neutralist theory *as it is commonly perceived* is represented in Figure 10-8. All genotypes are shown as being closely packed about the base line, D/M = 1.00. The genotypes differ neither under different situations in the population arena nor, on the average, at different times in the time arena. This view corresponds to the formal definition

151

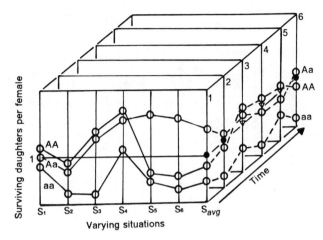

Figure 10-7. Population arenas arranged through time. The S_{avg} borders of successive population arenas form a plane, the *time arena*. Populations are assumed to be at equilibrium with respect to size; they may (solid circle) or may not (inverted triangle) be at equilibrium simultaneously with respect to the frequencies of A and a (see Figure 6-4). Evolutionary studies are mainly concerned with events occurring in the time arena; these studies deal with averages and, of necessity, fail to examine the norms of reaction that characterize each population arena. (From Wallace, 1989a, courtesy of *Quarterly Review of Biology*.)

of neutrality that was given at the start of this chapter: any allele can be substituted for any other in any cell of any individual under any situation without altering the subsequent development or eventual fitness of the altered individual.

Those who have been active in developing the neutralist theory (Kimura, 1983; Nei, 1987) are not this rigid in their view of neutrality; in their view a neutral allele is one that behaves as if it were neutral ("has behaved" would be more precise because no one has access to the future). Selection pressures that fluctuate through time or that vary depending upon the individual's situation result in apparent neutrality. Given, however, that the environment changes or that new alleles at other loci become common in the population, a once-neutral allele may become nonneutral.

The pragmatic view described above can be illustrated in one form as in Figure 10-9. Here neutrality is seen only as an average in each population arena; this average neutrality persists generation after gen-

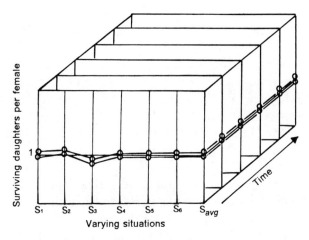

Figure 10-8. Extremely simple representations of population and time arenas resulting from gene neutrality in the strict sense where a definition of neutrality might read: two alleles are neutral if either one could be exchanged for the other in any cell, tissue, or organ of any individual within any environment without affecting the subsequent development or eventual fitness of the individual. (From Wallace, 1989a, courtesy of *Quarterly Review of Biology*.)

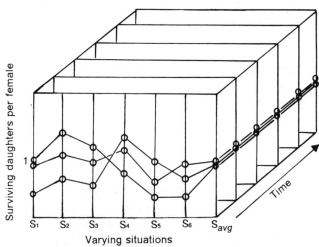

Figure 10-9. The apparent neutrality of alleles when viewed in the time arena, and the nonneutrality of the same alleles revealed by the evolutionarily significant reaction norms of the three genotypes—norms that lie within the population arena. To the extent that the apparent neutrality results from averaging over many situations (S_1, S_2, . . .), the removal of certain situations or the origin of new ones might alter circumstances considerably. This diagram and those shown in Figures 10-7 and 10-8 illustrate the gamut of reasonable views concerning selective neutrality.

eration. Within each population arena, however, the average fitnesses of different genotypes under different situations vary. Thus, as the neutralists claim, altering the population arena by removing certain situations or by adding new ones may cause neutral alleles to become nonneutral. Furthermore, because situations refer to background genotypes as well as to environmental conditions, the removal from the population of certain alleles (from any locus) or a sudden increase in the frequency of others can disturb the neutrality of a given allele (see Mayr, 1954).

A second version of the pragmatic view of neutrality can be represented as in the diagram illustrated in Figure 10-7. The future can only be surmised; the past, however, can be assessed. Viewing the time arena from a position near the arrow's head, one can judge the effective neutrality of alleles by their average fitnesses as revealed in the S_{avg} margin of each population arena. The alleles represented in Figure 10-7 are obviously not without an average effect on their carriers (*aa* individuals in the diagram have consistently exhibited low fitness); however, the overall view may under other circumstances lead one to conclude that the two alleles are effectively neutral.

Thus, it appears that a great deal of the seeming disagreement between neutralists and selectionists hinges on perceptions of the living world. Strict adherence to a view such as that shown in Figure 10-8 leads to the view that evolution can proceed only under novel environmental situations that call for drastically altered genetic responses, and only when newly arisen (favorable) mutations make these responses possible (Nei, 1987:427). Under this view, evolutionary rates are constrained by the rate at which favorable mutations arise.

A view of the living world such as that illustrated in Figure 10-9 suggests that existing genetic variation—even seemingly neutral variation—can respond to many (most, if not all) evolutionary challenges. This is so because neutrality is a reflection of average fitnesses within the population arena; dissimilar norms of reaction reveal, however, that novel situations are likely to evoke new fitness relationships between presently neutral alleles.

The most extreme view is, of course, that shown in Figure 10-7, in which even the *average* fitnesses of individuals of different genotypes differ considerably, and neutrality becomes an impression gained by looking back through successive averages that have fluctuated without one genotype or the other being systematically favored. That view, as was emphasized earlier, is an assessment of the past, an assessment that permits us only to guess at the future.

In an early discussion (Wallace, 1989a) similar to that presented in the preceding paragraphs, I posed a rhetorical question that can be paraphrased as follows: Might there not be forces operating within population arenas that tend to compress the (average) points which appear in the time arena (as shown in Figure 10-9), thus giving an impression of neutrality? I believe there are such forces; indeed, they were suggested in the personal remarks appended to the introductory chapter.

Fitness is an extremely complex attribute of an individual; it is a composite of many components, each of which in turn is a composite of numerous subcomponents. The number of hierarchies of components and subcomponents is certainly much greater than two; they extend downward until they encompass the actions of and interactions between individual genes. Many persons attempt to impose simplicity on these complexities by saying that the individual is the unit of selection. John Maynard Smith (1989) has gone even further by saying that only collections of individuals (genotypes, for example) have fitnesses; individuals, in his view, do not. I believe the position taken by Maynard Smith is self-contradictory for most higher organisms because each individual possesses a unique genotype; hence, dividing a population into smaller and smaller genotypic classes eventually divides it into individuals.

The more complex any attribute is, the smaller the correlation between it and any of its components. Take, for example, several columns of random digits and the sums of the individual rows:

					Sum
5	3	6	7	6	27
7	0	6	6	3	22
0	3	8	7	1	19
7	4	6	4	5	26
5	8	4	4	6	27
9	5	7	6	6	33
0	3	1	9	2	15
2	6	2	1	6	17
4	1	4	6	7	22
9	7	1	7	6	30

Fifty Years of Genetic Load

The expected correlation (r) between the individual digits in any column and the sum of all (n) columns can be shown to be

$$r = \sqrt{\frac{1}{n}}$$

Hence, if n were 100, the expected correlation would be 0.10. Four hundred observations would be needed for such a small correlation to be significant at the 5% level; nearly 1000 would be required to reduce the level of significance to 1%. An experimentalist might be justified in concluding that the sums are unrelated to the digits in any one column although each column is as important as any other.

The survival of individuals in natural populations is analogous in many respects to victory in a sporting event. Consider horse racing. Many physical attributes of a horse contribute to its speed and stamina over a specified distance. Nevertheless, the rule used by race horse breeders is "breed speed with speed." No breeding program that is based on a single (or few) aspects of the animal's phenotype is likely to result in faster horses (see Gaffney and Cunningham, 1988; Hill, 1988).

Someone analyzing traits of obvious importance with respect to an athlete's ability (for example, a boxer's) would encounter traits that themselves are correlated, one with the other (circumference of biceps and that of forearm or wrist). In that case, the correlation of the sum of n elements with any individual element equals

$$r = \sqrt{\frac{1 - \gamma}{n} + \gamma}$$

where γ is the correlation among elements. If all elements are perfectly correlated—for example, $r = \sqrt{1}$, or 1—the sum is perfectly correlated with any element. If $\gamma = 0$, $r = \sqrt{1/n}$, as cited in the case of random (uncorrelated) digits.

The equation

$$r = \sqrt{\frac{1 - \gamma}{n} + \gamma}$$

Still to Come . . .

can be solved for $r = 0$:

$$1 - \gamma + n\gamma = 0$$
$$\gamma = -\frac{1}{n-1}.$$

This equation was cited in Chapter 1 in reference to terminology: Does a slight negative correlation among recognizably important components of fitness suffice to make them "neutral"? In these concluding paragraphs, however, this equation is discussed with reference to several points that have been made in later chapters of this book.

The absence of correlation between a component of a complex trait and the trait itself is a characteristic only of the upper tail of a large distribution of individuals that otherwise tend to be overlooked: there are no legless race horses, nor are there armless prize fighters. The individuals among whom one finds little or no correlation between individual components and the inclusive trait (such as fitness) are those that have survived an extended culling process—a process I call rank-order (soft) selection. (The neutral theory of faculty composition that I mentioned whimsically in the introductory chapter (p. 3) can now be seen to have as its bases, first, the intense culling that occurs during the elementary, undergraduate, graduate, and postgraduate education of college and university professors, and, second, the negative correlations that exist among any specialist's potential areas of expertise. Despite the heroic efforts of university search committees, candidates for professorships are essentially equally qualified overall, and the eventual choice becomes virtually a matter of chance—that is, within the limited pool of adequately trained candidates.)

Individuals that survive the culling process referred to above are those that establish conditions affecting the population arena and the evolutionarily important norms of reaction falling within that arena. Norms of reaction falling outside the population area, while they may have instructional value, have little or no bearing on the pattern of selection in natural populations. Dobzhansky and M. L. Queal (1938) and Dobzhansky et al. (1942) showed rather convincingly that in *D. pseudoobscura*, subvitality, fertility, and speed of development are correlated: flies with poor viability are often sterile and, more often than not, develop more slowly than flies with quasi-normal viabilities. Because these wretched flies would not normally survive in a natural population, neither their sterility nor the time needed for their development is of any importance. Much more important is the negative correlation that probably holds within the population arena: a reduction of only a few hours in development time (an enhancer of fitness) can be achieved only at the expense of final size and, in females, fewer eggs (items that lower fitness). Alan Robertson (1955) pointed out that if any component of fitness can be enhanced by artificial selection, either the population from which the experimental material was obtained was not at equilibrium, or there exists a negative correlation among components of fitness. I now add the caveat that these negative correlations need exist only within the population arena, and involve only those individuals who have survived and who perpetuate the population (see Figure 10-10).

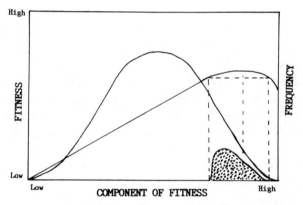

Figure 10-10. Even nonneutral phenotypic traits (components of fitness) become neutral in populations at equilibrium. The horizontal axis represents a component of fitness (fecundity, speed of development, or [in prize fighting] a boxer's reach); the bell-shaped curve represents the distribution of this component among all individuals (vertical axis, right). The vertical axis (left) represents total fitness. Over much of its range, the component of fitness is directly correlated with total fitness; at some point, however, the component exhibits an optimum value after which it is negatively correlated with fitness (enormous numbers of eggs can be produced only at the expense of other fitness components; grotesquely long arms hinder rather than help a boxer). Rank-order selection proceeds from the bottom upward in this diagram, the eventual survivors lie at and near the highest point of the fitness curve. Hence, these survivors fall more or less equally on the two sides of the component's optimum value, thus making the trait (among survivors) essentially neutral. (From Reeve et al., 1990.)

Still to Come ...

Negative correlations between components of fitness within the population arena, causing these individual components to be *non*-correlated with overall fitness, provide the compressive forces that lead to an appearance of neutrality in the time arena (as in Figure 10-9). If that point is clear, and if my analysis has been correct, then my interest in the term *neutrality* all but vanishes. As I have said earlier, my interest is in the situation, in the "reality" as far as reality can be discerned; it is not in the words used by various persons in their attempts to describe a complex situation. It remains the task of the next generation of population biologists to decide whether the apparent neutrality of selectively important genetic factors and phenotypic traits demonstrates (1) the essential correctness of the neutralist view, (2) the essential correctness of the selectionist view, or (3) an improper view of a population's biology that has entrapped many persons in needless debate. These are matters whose resolution is yet to come.

Personal comments

Most people are somewhat irrational. During the period of William Shockley's notoriety concerning his views on racial IQs, student associations at several colleges and universities invited him to speak, even though his arguments were transparently faulty. I once took time to view a motion picture which supposedly demonstrated the transformation of characters from one breed of ducks to another following injections of alien DNA. A more rational colleague refused to attend the showing because the movie would have had no effect upon his extreme skepticism concerning this research.

A scientific controversy arises as a difference of opinion that cannot be resolved by available technologies or systems of logic. The controversy grows in intensity and notoriety as persons mindlessly repeat experiments and observations already recognized as inadequate for the task at hand. In my opinion, the eventual outcome of a scientific controversy, unfortunately, is often to provide a topic of animated conversation for those not immediately or personally concerned.

Population geneticists have long had their differences over genetic loads and, later, over the neutralist theory. I take pride in my contributions to these debates. I take equal pride, however, that after these many years I can receive reviewers' comments such as the following: "Bruce Wallace has been the unofficial gadfly of population genetics for decades. His ideas have been controversial and idiosyncratic, and outside the main stream. But they have also been imaginative and stimulating, both of thought and research. This article follows the Wallace tradition. Bruce and I have been friendly, respectful opponents on several issues in the past, and I won't break the tradition now . . . [signed James F. Crow]."

One person's controversy can be another person's—or two persons'—intellectual challenge. If so, it can be a delight!

REFERENCES

Ayala, F. J. 1966. Evolution of fitness. I. Improvement in the productivity and size of irradiated populations of *Drosophila serrata* and *D. birchii*. Genetics 53: 883–895.
Ayala, F. J. 1968. Genotype, environment, and population numbers. Science 162:1453–1459.
Ayala, F. J. 1969. Genetic polymorphism and interspecific competitive ability in *Drosophila*. Genet. Res. 14:95–102.
Ayala, F. J., M. E. Gilpin, and J. G. Ehrenfeld. 1973. Competition between species: Theoretical models and experimental tests. Theor. Popul. Biol. 4:331–356.
Beatty, J. 1987. Weighing the risks: Stalemate in the classical/balance controversy. J. Hist. Biol. 20:289–319.
Blaylock, B. G., and H. H. Shugart, Jr. 1972. The effect of radiation-induced mutations on the fitness of *Drosophila* populations. Genetics 72:469–474.
Bøggild, O., and J. Keiding. 1958. Competition in house fly larvae, experiments involving a DDT-resistant and a susceptible strain. Oikos 9:1–25.
Bonnier, G., U.-B. Jonsson, and C. Ramel. 1958. Selection pressure on irradiated populations of *Drosophila melanogaster*. Hereditas 44:378–406.
Bray, J. R. 1956. Gap phase replacement in a maple-basswood forest. Ecology 37:598–600.
Britten, R. J., and E. H. Davidson. 1969. Gene regulation for higher cells: A theory. Science 165:349–357.
Carson, H. L. 1957. Production of biomass as a measure of fitness of experimental populations of *Drosophila*. Genetics 42:363–364.
Carson, H. L. 1958. Increase in fitness in experimental populations resulting from heterosis. Proc. Natl. Acad. Sci. USA 44:1136–1141.
Carson, H. L. 1961. Heterosis and fitness in experimental populations of *Drosophila melanogaster*. Evolution 15:496–509.

References

Clarke, B. 1973a. The effect of mutation on population size. Nature (Lond.) 242:196–197.
Clarke, B. 1973b. Mutation and population size. Heredity 31:367–379.
Cooper, W. S., and R. H. Kaplan. 1982. Adaptive "coin-flipping": A decision-theoretic examination of natural selection for random individual variation. J. Theor. Biol. 94:135–151.
Coyne, J. A. 1974. The evolutionary origin of hybrid inviability. Evolution 28:505–506.
Crick, F. 1971. General model for the chromosomes of higher organisms. Nature (Lond.) 234:25–27.
Crow, J. F. 1948. Alternate hypotheses of hybrid vigor. Genetics 33:477–487.
Crow, J. F. 1958. Some possibilities for measuring selection intensities in man. Hum. Biol. 30:1–13.
Crow, J. F. 1987. Muller, Dobzhansky, and overdominance. J. Hist. Biol. 20:351–380.
Crow, J. F., and R. G. Temin. 1964. Evidence for the partial dominance of recessive lethal genes in natural populations of *Drosophila*. Am. Nat. 98:21–33.
Davey, R. B., and D. C. Reanney. 1980. Extrachromosomal genetic elements and the adaptive evolution of bacteria. Evol. Biol. 13:113–147.
Dawkins, R. 1976. *The Selfish Gene.* New York: Oxford University Press.
Dobzhansky, Th. 1937. *Genetics and the Origin of Species.* New York: Columbia University Press.
Dobzhansky, Th. 1947. Genetics of natural populations. XIV. A response of certain gene arrangements in the third chromosome of *Drosophila pseudoobscura* to natural selection. Genetics 32:142–160.
Dobzhansky, Th. 1948. Genetics of natural populations. XVIII. Experiments on chromosomes of *Drosophila pseudoobscura* from different geographic regions. Genetics 33:588–602.
Dobzhansky, Th. 1950. Genetics of natural populations. XIX. Origin of heterosis through natural selection in populations of *Drosophila pseudoobscura*. Genetics 35:288–302.
Dobzhansky, Th. 1952. Nature and origin of heterosis. In J. W. Gowen, ed., *Heterosis,* pp. 218–223. Ames: Iowa State College Press.
Dobzhansky, Th. 1968. Adaptedness and fitness. In R. C. Lewontin, ed. *Population Biology and Evolution,* pp. 109–121. Syracuse: Syracuse University Press.
Dobzhansky, Th. 1970. *Genetics of the Evolutionary Process.* New York: Columbia University Press.
Dobzhansky, Th., and C. Epling. 1944. Contributions to the genetics, taxonomy, and ecology of *Drosophila pseudoobscura* and its relatives. Carnegie Inst. Wash. Publ. 554:1–183.
Dobzhansky, Th., A. M. Holz, and B. Spassky. 1942. Genetics of natural populations. VIII. Concealed variability in the second and fourth chromosomes of *Drosophila pseudoobscura* and its bearing on the problem of heterosis. Genetics 27:463–490.
Dobzhansky, Th., and O. Pavlovsky. 1961. A further study of fitness of chromo-

References

somally polymorphic and monomorphic populations of *Drosophila pseudoobscura*. Heredity 16:169–179.
Dobzhansky, Th., and M. L. Queal. 1938. Genetics of natural populations. II. Genetic variation in populations of *Drosophila pseudoobscura* inhabiting isolated mountain ranges. Genetics 23:463–484.
Dobzhansky, Th., and B. Spassky. 1944. Genetics of natural populations. XI. Manifestation of genetic variants in *Drosophila pseudoobscura* in different environments. Genetics 29:270–290.
Dubinin, N. P. 1964. *Problems of Radiation Genetics* (trans. G. H. Beale). Edinburgh: Oliver and Boyd.
Engebrecht, J., M. Simon, and M. Silverman. 1985. Measuring gene expression with light. Science 227:1345–1347.
Feller, W. 1950. *An Introduction to Probability Theory and Its Applications*. New York: John Wiley and Sons.
Feller, W. 1967. On fitness and the cost of natural selection. Genet. Res. 9:1–15.
Fisher, R. A. 1930. *The Genetical Theory of Natural Selection*. Oxford: Clarendon Press. [1958. 2d ed. New York: Dover Press].
Gaffney, B., and E. P. Cunningham. 1988. Estimation of genetic trends in racing performance of thoroughbred horses. Nature 332:722–724.
Garton, D. W., R. K. Koehn, and T. M. Scott. 1984. Multiple-locus heterozygosity and the physiological energetics of growth in the coot clam, *Malinia lateralis*, from a natural population. Genetics 108:445–455.
Golenberg, Ed. 1989. Population Ecology of Individuals, by Adam Łomnicki (a review). Q. Rev. Biol. 64:88–89.
Gould, S. J., and N. Eldredge. 1977. Punctuated equilibria: The tempo and mode of evolution reconsidered. Paleobiology 3:115–151.
Govind, C. K. 1989. Asymmetry in lobster claws. Am. Sci. 77:468–474.
Govind, C. K., and J. Pearce. 1986. Differential reflex activity determines claw and closer muscle asymmetry in developing lobsters. Science 233:354–356.
Gross, W. G., P. B. Siegel, R. W. Hall, C. H. Domermuth, and R. T. DuBoise. 1980. Production and persistence of antibodies in chickens to sheep erythrocytes. 2. Resistance to infectious diseases. Poult. Sci. 59:205–210.
Gupta, A. P., and R. C. Lewontin. 1982. A study of reaction norms in natural populations of *Drosophila pseudoobscura*. Evolution 36:934–948.
Haldane, J. B. S. 1924. A mathematical theory of natural and artificial selection. Part I. Trans. Camb. Philos. Soc. 23:19–41.
Haldane, J. B. S. 1932. *The Causes of Evolution*. London: Longmans, Green and Co. [Reissued by Cornell University Press, 1966.]
Haldane, J. B. S. 1937. The effect of variation on fitness. Am. Nat. 71:337–349.
Haldane, J. B. S. 1953. Animal populations and their regulation. New Biol. 15:9–24.
Haldane, J. B. S. 1957. The cost of natural selection. J. Genet. 55:511–524.
Haldane, J. B. S. 1960. More precise expressions for the cost of natural selection. J. Genet. 57:351–360.
Hamilton, W. D. 1964a. The genetical evolution of social behavior. I. J. Theor. Biol. 7:1–16.

References

Hamilton, W. D. 1964b. The genetical evolution of social behavior. II. J. Theor. Biol. 7:17–52.

Harper, J. L. 1977. *Population Biology of Plants*. New York: Academic Press.

Hartl, D. L., and H. Jungen, 1979. Estimation of average fitness of populations of *Drosophila melanogaster* and the evolution of fitness in experimental populations. Evolution 33:371–380.

Hill, W. G. 1988. Selective breeding: Why aren't horses faster? Nature (Lond.) 332:678.

Hiraizumi, Y. 1965. Effects of X-ray induced mutations on several components of fitness. Ann. Rep. Natl. Inst. Genet. (Japan) 15:109–111.

Hollingshead, L. 1930. A lethal factor in *Crepis* effective only in an interspecific hybrid. Genetics 15:114–140.

Jungen, H., and D. L. Hartl. 1979. Average fitness of populations of *Drosophila melanogaster* as estimated using compound-autosome strains. Evolution 33:359–370.

Katznelson, J. 1976. Domestication processes in *Trifolium berytheum* Boiss. In S. Karlin and E. Nevo, eds., *Population Genetics and Ecology*, pp. 91–103. New York: Academic Press.

Kidwell, M. G., J. F. Kidwell, and J. A. Sved. 1977. Hybrid dysgenesis in *Drosophila melanogaster*: A syndrome of aberrant traits including mutation, sterility and male recombination. Genetics 86:813–833.

Kimura, M. 1968. Evolutionary rate at the molecular level. Nature (Lond.) 217:624–626.

Kimura, M. 1983. *The Neutral Theory of Molecular Evolution*. Cambridge: Cambridge University Press.

Kimura, M., and J. F. Crow. 1964. The number of alleles that can be maintained in a finite population. Genetics 49:725–738.

Kimura, M., and J. F. Crow. 1969. Natural selection and gene substitution. Genet. Res. 13:127–141.

Kimura, M., and T. Ohta. 1971. *Theoretical Aspects of Population Genetics*. Princeton: Princeton University Press.

King, J. L., and T. H. Jukes. 1969. Non-Darwinian evolution. Science 164:788–798.

Lerner, I. M. 1954. *Genetic Homeostasis*. Edinburgh: Oliver and Boyd.

Levins, R. 1968. *Evolution in Changing Environments*. Princeton: Princeton University Press.

Lewontin, R. C. 1968. Introduction. In R. C. Lewontin, ed., *Population Biology and Evolution*, pp. 1–4. Syracuse: Syracuse University Press.

Lewontin, R. C. 1974. *The Genetic Basis of Evolutionary Change*. New York: Columbia University Press.

Lewontin, R. C., and J. L. Hubby. 1966. A molecular approach to the study of genic heterozygosity in natural populations. 2. Amount of variation and degree of heterozygosity in natural populations of *Drosophila pseudoobscura*. Genetics 54:595–609.

Linck, A. J. 1961. The morphological development of the fruit of *Pisum sativum*, var. *Alaska*. Phytomorphology 11: 79–84.

References

Łomnicki, A. 1988. *Population Ecology of Individuals.* Princeton: Princeton University Press.

MacIntyre, R. J. 1982. Regulatory genes and adaptation: Past, present, and future. Evol. Biol. 15:247–285.

Maruyama, T., and J. F. Crow. 1975. Heterozygous effects of X-ray induced mutations on viability of *Drosophila melanogaster.* Mutat. Res. 27:241–248.

Mather, K. 1953. Genetical control of stability in development. Heredity 7:297–336.

Maynard Smith, J. 1982. *Evolution and the Theory of Games.* Cambridge: Cambridge University Press.

Maynard Smith, J. 1989. *Evolutionary Genetics.* Oxford: Oxford University Press.

Mayr, E. 1942. *Systematics and the Origin of Species.* New York: Columbia University Press.

Mayr, E. 1954. Change of genetic environment and evolution. In J. Huxley, A. C. Hardy, and E. B. Ford, eds., *Evolution as a Process,* pp. 157–180. London: Allen and Unwin.

Mayr, E. 1982. *The Growth of Biological Thought: Diversity, Evolution, and Inheritance.* Cambridge: Belknap Press of Harvard University Press.

Mertz, D. B. 1988. Population Ecology of Individuals, by Adam Łomnicki (a review). Science 241:478.

Moser, H. 1958. The dynamics of bacterial populations maintained in the chemostat. Carnegie Inst. Wash. Publ. 614:1–136.

Mukai, T. 1985. Experimental verification of the neutral theory. In T. Ohta and K. Aoki, eds., *Population Genetics and Molecular Evolution,* pp. 125–145. Tokyo: Japan Scientific Society Press.

Mukai, T., H. E. Schaffer, and C. C. Cockerham. 1972. Genetic consequences of truncation selection at the phenotypic level in *Drosophila melanogaster.* Genetics 72:763–769.

Mukai, T., and I. Yoshikawa. 1964. Heterozygous effects of radiation-induced mutations on viability in homozygous and heterozygous genetic backgrounds in *Drosophila melanogaster.* Jpn. J. Genet. 38:282–287.

Muller, H. J. 1927. Artificial transmutation of the gene. Science 66:84–87.

Muller, H. J. 1950. Our load of mutations. Am. J. Hum. Genet. 2:111–176.

Muller, H. J. 1959. The mutation theory reexamined. Proc. 10th Int. Congr. Genet. 1:306–317.

Nabours, R. K., and L. L. Kingsley. 1934. The operations of a lethal factor in *Apotettix eurycephalus* (grouse locusts). Genetics 19:323–328.

National Academy of Sciences, National Research Council. 1956. The Biological Effects of Atomic Radiation. Washington, D.C.: National Academy Press.

Neel, J. V., and W. J. Schull. 1956. The effect of exposure to the atomic bombs on pregnancy termination in Hiroshima and Nagasaki. Natl. Acad. Sci., Natl. Res. Council Publ. 461.

Nei, M. 1987. *Molecular Evolutionary Genetics.* New York: Columbia University Press.

Nöthel, Horst. 1987. Adaptation of *Drosophila melanogaster* populations to high mutation pressure: Evolutionary readjustment of mutation rates. Proc. Natl. Acad. Sci. USA 84:1045–1049.

O'Brien, S. J., D. E. Wildt, D. Goldman, C. R. Merril, and M. Bush. 1983. The cheetah is depauperate in genetic variation. Science 221:459–462.

Ow, D. W., K. V. Wood, M. DeLuca, J. R. deWet, D. R. Helinski, and S. H. Howell. 1986. Transient and stable expression of the firefly luciferase gene in plant cells and transgenic plants. Science 234:856–859.

Paget, O. E. 1954. A cytological analysis of irradiated populations. Am. Nat. 88:105–107.

Paquin, C., and J. Adams. 1983. Frequency of fixation of adaptive mutations is higher in evolving diploid than haploid yeast populations. Nature (Lond.) 302:495–500.

Penrose, L. S. 1949. The meaning of "fitness" in human populations. Ann. Eugen. 14:301–304.

Prout, T. 1954. Genetic drift in irradiated experimental populations of *Drosophila melanogaster*. Genetics 39:529–545.

Reeve, R., E. Smith, and B. Wallace. 1990. Components of fitness become effectively neutral in equilibrium populations. Proc. Natl. Acad. Sci. USA 87:2018–2020.

Robertson, A. 1955. Selection in animals: Synthesis. Cold Spring Harbor Symp. Quant. Biol. 20:225–229.

Russell, W. L. 1962. (See U.N. Scientific Committee on the Effects of Atomic Radiation.)

Schauer, A., M. Ranes, R. Santamaria, J. Guijarro, E. Lawlor, C. Mendez, K. Chater, and R. Losick. 1988. Visualizing gene expression in time and space in the filamentous bacterium *Streptomyces coelicolor*. Science 240:768–772.

Schmalhausen, I. I. 1949. *Factors of Evolution*. Philadelphia: Blakiston. [Reissued by University of Chicago Press, 1986.]

Simmons, M. J., and J. F. Crow. 1977. Mutations affecting fitness in *Drosophila* populations. Ann. Rev. Genet. 11:49–78.

Simpson, G. G. 1944. *Tempo and Mode in Evolution*. New York: Columbia University Press.

Singh, S. M., and E. Zouros. 1978. Genetic variation associated with growth rate in the American oyster (*Crassostrea virginica*). Evolution 32:342–353.

Sober, E., and R. C. Lewontin. 1984. Artifact, cause and genic selection. In R. N. Brandon and R. M. Burian, eds., *Genes, Organisms, Populations*, pp. 109–132. Cambridge, Mass.: MIT Press.

Stebbins, G. L. 1950. *Variation and Evolution in Plants*. New York: Columbia University Press.

Stebbins, G. L. 1958. Longevity, habitat, and release of genetic variability in the higher plants. Cold Spring Harbor Symp. Quant. Biol. 23:365–378.

Stephens, S. G. 1946. The genetics of "Corky." I. The New World alleles and their possible role as an interspecific isolating mechanism. J. Genet. 47:150–161.

Stephens, S. G. 1950. The genetics of "Corky." II. Further studies on its genetic basis in relation to the general problem of interspecific isolating mechanisms. J. Genet. 50:9–20.

References

Stern, C. G. Carson, M. Kinst, E. Novitski, and D. Uphoff. 1952. The viability of heterozygotes for lethals. Genetics 37:413–449.
Sved, J. A., T. E. Reed, and W. F. Bodmer. 1967. The number of balanced polymorphisms that can be maintained in a natural population. Genetics 55: 469–481.
Thoday, J. M. 1953. Components of fitness. Symp. Soc. Exp. Biol. 7:96–113.
Thompson, V. 1985. Documentation of the electrophoresis revolution in Drosophila population genetics. Dros. Info. Serv. 61:8.
Thompson, V. 1986. Half-chromosome viability and synthetic lethality in Drosophila melanogaster. J. Hered. 77:385–388.
Turelli, M., and L. R. Ginzburg. 1983. Should individual fitness increase with heterozygosity? Genetics 104:191–209.
Turner, J. R. G., and M. H. Williamson. 1968. Population size, natural selection, and genetic load. Nature (Lond.) 218:700.
U.N. Scientific Committee on the Effects of Atomic Radiation. 1962. *Report of the U.N. Scientific Committee on the Effects of Atomic Radiation*. New York: United Nations.
Vann, E. G. 1966. The fate of X-ray induced chromosomal rearrangements introduced into laboratory populations of Drosophila melanogaster. Am. Nat. 100: 425–449.
Van Valen, L. 1975. Life, death, and energy of a tree. Biotropica 7:259–269.
Waddington, C. H. 1953. Genetic assimilation of an acquired character. Evolution 7:118–126.
Waddington, C. H. 1956. Genetic assimilation of the *bithorax* phenotype. Evolution 10:1–13.
Wallace, B. 1952. The estimation of adaptive values of experimental populations. Evolution 6:333–341.
Wallace, B. 1956. Studies on irradiated populations of Drosophila melanogaster. J. Genet. 54:280–293.
Wallace, B. 1958. The average effect of radiation-induced mutations on viability in Drosophila melanogaster. Evolution 12:532–536.
Wallace, B. 1959a. Studies of the relative fitnesses of experimental populations of Drosophila melanogaster. Am. Nat. 93:295–314.
Wallace, B. 1959b. The role of heterozygosity in Drosophila populations. Proc. 10th Int. Congr. Genet. 1:408–419.
Wallace, B. 1963a. A comparison of the viability effects of chromosomes in heterozygous and homozygous condition. Proc. Natl. Acad. Sci. USA 49:801–806.
Wallace, B. 1963b. Modes of reproduction and their genetic consequences. In W. D. Hanson and H. F. Robinson, eds., *Statistical Genetics and Plant Breeding*, pp. 3–20. Natl. Acad. Sci., Natl. Res. Council Publ. 982.
Wallace, B. 1963c. The annual invitation lecture. Genetic diversity, genetic uniformity, and heterosis. Can. J. Genet. Cytol. 5:239–253.
Wallace, B. 1965. The viability effects of spontaneous mutations in Drosophila melanogaster. Am. Nat. 99:335–348.
Wallace, B. 1968a. *Topics in Population Genetics*. New York: Norton.

References

Wallace, B. 1968b. Polymorphism, population size, and genetic load. In R. C. Lewontin, ed., *Population Biology and Evolution*, pp. 87–108. Syracuse: Syracuse University Press.

Wallace, B. 1975a. Hard and soft selection revisited. Evolution 29:465–473.

Wallace, B. 1975b. Gene control mechanisms and their possible bearing on the neutralist-selectionist controversy. Evolution 29: 193–202.

Wallace, B. 1976. The structure of gene control regions and its bearing on diverse aspects of population genetics. In S. Karlin and E. Nevo, eds., *Population Genetics and Ecology*, pp. 499–521. New York: Academic Press.

Wallace, B. 1979. Population size, environment, and the maintenance of laboratory cultures of *Drosophila melanogaster*. Genetika 10:9–16.

Wallace, B. 1981. *Basic Population Genetics*. New York: Columbia University Press.

Wallace, B. 1982. Phenotypic variation with respect to fitness: The basis of rank-order selection. Biol. J. Linn. Soc. 17:269–274.

Wallace, B. 1983. A possible explanation for observed differences in the geographical distributions of chromosomal rearrangements of plants and *Drosophila*. Egypt. J. Genet. Cytol. 13:121–136.

Wallace, B. 1986a. Genetic change-over in *Drosophila* populations. Proc. Natl. Acad. Sci. USA 83:1374–1378.

Wallace, B. 1986b. On the viability effects of chromosomes in *Drosophila*. Am. Nat. 128:272–281.

Wallace, B. 1987a. Analyses of genetic change-over in *Drosophila* populations. Z. Zool. Syst. Evolutionsforsch. 25:40–50.

Wallace, B. 1987b. Ritualistic combat and allometry. Am. Nat. 129:775–776.

Wallace, B. 1987c. Evolution through group selection, by V. C. Wynne-Edwards (a review). Bull. Math. Biol. 49:629–632.

Wallace, B. 1987d. The Wilhelmine E. Key Invitational Lecture: Fifty years of genetic load. J. Hered. 78:134–142.

Wallace, B. 1988. Selection for the inviability of sterile hybrids. J. Hered. 79:204–210.

Wallace, B. 1989a. One selectionist's perspective. Q. Rev. Biol. 64:127–145.

Wallace, B. 1989b. Analyzing variation in egg-to-adult viability in experimental populations of *Drosophila melanogaster*. Proc. Natl. Acad. Sci. USA 86:2117–2121.

Wallace, B., and C. E. Blohowiak. 1985a. Rank-order selection and the analysis of data obtained by *ClB*-like procedures. Evol. Biol. 19:99–146.

Wallace, B., and C. E. Blohowiak. 1985b. Rank-order selection and the interpretation of data obtained by *ClB*-like procedures. Biol. Zentralbl. 104:683–700.

Wallace, B., and T. L. Kass. 1974. On the structure of gene control regions. Genetics 77:541–548.

Wallace, B., and J. C. King. 1952. A genetic analysis of the adaptive values of populations. Proc. Natl. Acad. Sci. USA 38:706–715.

Wallace, B., and M. Vetukhiv. 1955. Adaptive organization of the gene pools of *Drosophila* populations. Cold Spring Harbor Symp. Quant. Biol. 20:303–310.

Westoby, M. 1981. How diversified seed germination behaviour is selected. Am. Nat. 118:882–885.

References

Wilson, E. O. 1975. *Sociobiology: The New Synthesis.* Cambridge: Belknap Press of Harvard University Press.
Wood, K. V., Y. A. Lam, H. H. Seliger, and W. D. McElroy. 1989. Complementary DNA coding click beetle luciferases can elicit bioluminescence of different colors. Science 244:700–702.
Woodruff, R. C., and J. N. Thompson, Jr. 1980. Hybrid release of mutator activity and the genetic structure of natural populations. Evol. Biol. 12:129–162.
Wourms, J. P. 1972. The developmental biology of annual fishes. III. Preembryonic and embryonic diapause of variable duration in the eggs of annual fishes. J. Exp. Zool. 182:389–414.
Wright, S. 1931. Evolution in Mendelian populations. Genetics 16:97–159.
Wright, S., and Th. Dobzhansky. 1946. Genetics of natural populations. XII. Experimental reproduction of some of the changes caused by natural selection in certain populations of *Drosophila pseudoobscura.* Genetics 31:125–156.
Wright, T. F. R. 1963. The genetics of an esterase in *Drosophila melanogaster.* Genetics 48:787–801.
Wynne-Edwards, V. C. 1962. *Animal Dispersion in Relation to Social Behaviour.* Edinburgh: Oliver and Boyd.
Wynne-Edwards, V. C. 1987. *Evolution through Group Selection.* Oxford: Blackwell Scientific Publications.
Zouros, E., S. M. Singh, and H. E. Miles. 1980. Growth rate in oysters: An overdominant phenotype and its possible explanations. Evolution 34:856–867.

INDEX

Acer saccharinum
  competition between seedlings, 92–93
  one-fruited mutant, 122
Adams, J., 59
Adaptedness, 105
Age as contributor to nongenetic variation, 16–17, 113–14
Apotettix eurycephalus, 15
Ayala, F. J., 83, 106

Balanced (genetic) load, 15
Beatty, J., 131
Blaylock, B. G., 106
Blohowiak, C. E., 35, 40, 102
Bøggild, O., 112
Bonnier, G., 43
Brassica campestris, 111
Britten, R. J., 61
Britten-Davidson model of gene control, 61

Cainism, 124
Carson, H. L., 106
Clarke, B., 83
ClB-like techniques, 24–28
Coadaptation, 4
Compound-autosome strains of *Drosophila melanogaster*, 106
Cooper, W. S., 144
Corky gene in New World cotton, 93–100

Cost of natural selection, 74–79
Coyne, J. A., 96
Crick, F., 61
Crow, J. F., 8, 12–15, 19, 41, 50, 56, 69–70, 78, 83, 100, 131
Cunningham, E. P., 156
Cy L tests
  heterozygous chromosomal combinations, 28
  homozygous chromosomal combinations, 26–27
  interpretation of data obtained by means of, 28–35
Cy L–Pm tests, 31–35

Darwinian fitness(es), average of ($\overline{W}$), 8–9
Davey, R. B., 110
Davidson, E. H., 61
Dawkins, R., 109
Demerec, M., 20
Density-dependent selection, 81–85
Dobzhansky, Th., 1, 4, 15–17, 25, 35, 105–106, 157
Drift (genetic) load, 16
Drosophila melanogaster
  compound-autosome strains, 106
  irradiated populations of, 24, 35–43
Dubinin, N. P., 21
Dykhuizen-Hartl effect, 103
Dysmetric (genetic) load, 16

Index

Eldredge, N., 46
Engebrecht, J., 64
Esterase-6 locus, first observed allozyme polymorphism, 69
Euterpe globosa, 114
Evolution of a university (an analogy), 2–3

Faculty composition, neutral theory of (an analogy), 3, 157
Feller, W., 77
Fisher, R. A., 1, 15, 69
Fitness
  inclusive, 12
  loss of, through gene mutation, 12–14
  total, lack of correlation with its components, 7, 155–158
Founder theory, 4
Frequency-dependent selection, 81–85

Gaffney, B., 156
Garton, D. W., 118
Genetic assimilation, 4
Genetic drift, 2
Genetic homeostasis, 4
Genetic load(s)
  balanced, 15
  classification of, 15
  defined, 14
  drift, 16
  dysmetric, 16
  estimated from mutation rates, 18–19
  incompatibility, 16
  migration, 16
  mutational, 15
  segregational, 15
  substitution, 16
Ginsberg, L. R., 119
Golenberg, E., 144
Gould, S. J., 46
Govind, C. K., 141
Gupta, A. P., 150

Haldane, J. B. S., 1–2, 8, 12, 15–16, 46, 74, 86, 101, 133
Hamilton, W. D., 12
Hard selection, 81
  and soft selection, contrasting consequences of, 85–91
Harper, J. L., 122
Hartl, D. L., 106
Herskowitz, I., 67

Heterozygote advantage, consequences of, 44
Heterozygotes, role in producing ill-adapted homozygotes, 10–12
Heterozygous advantage, ubiquitous (Kimura-Crow model), 71–73

Inclusive fitness, 12
Incompatibility (genetic) load, 16
Intrinsic rate of increase (r) as a measure of population fitness, 105–106

Jukes, T. H., 3
Jungen, H. E., 106

Kaplan, R. H., 144
Kass, T. L., 62, 119
Kaufman, B. P., 20
Keiding, J., 112
Kidwell, M. G., 65
Kimura, M., 2, 4, 62, 70, 78, 100–101, 152
King, J. C., 6
King, J. L., 3
Kingsley, L. L., 15

Lerner, I. M., 4, 62, 103
Levins, R., 144
Lewontin, R. C., 109, 141, 145, 150
Linck, A. J., 122
Łomnicki, A., 144, 149

Maruyama, T., 56, 69
Mather, K., 124
Maynard Smith, J., 144, 155
Mayr, E., 1, 4, 154
Mertz, D. B., 144
Microenvironments as contributors to nongenetic variation, 16–17, 114
Microevolution, 8
Migration (genetic) load, 16
Moser, H., 20
Mukai, T., 56, 65
Mulina lateralis, 118
Muller, H. J., 2, 6, 12, 15, 41, 46–49, 62, 66, 86
Mus musculus, 18
Mutation rate, estimated from genetic load, 17
Mutational (genetic) load, 15

Nabours, R. K., 15

Index

National Academy of Sciences, U.S., 20
Natural selection, cost of, 74–79
Neel, J. V., 18
Nei, M., 152, 154
Neutral theory of faculty composition (an analogy), 3, 157
Nongenetic variation, sources of
  age, 16–17, 113–114
  microenvironments, 16–17, 114
Norms of reaction, 150–151

O'Brien, S. J., 118
Ohta, T., 78, 101
Ow, D. W., 64

Paget, O., 134
Paquin, C., 59
Parental fertility, lowered by ill-adapted progeny, 10–14
Pavlovsky, O., 106
Pearce, J., 141
Penrose, L. S., 135
Persistence of populations, 104, 107–109
Phenotypic variation
  genetically based, classification of, 117
  proximate genetic basis, 121
  ultimate genetic basis, 121–125
Polymorphism, somatic, 122
Population, average fitness of ($\bar{W}$), 23–24
Population arena, 150
Population model
  heterozygotes are favored, 52–53
  homozygotes are favored, 50–52
Population regulation, Haldane's model of, 86–91
Prout, T., 45
Punctuated equilibrium, 46
Pyrophorus plagiophthalamus, 64

Queal, M. L., 157

Reanney, D. C., 110
Recessive lethals, partial dominance of, 41
Reeve, R., 7
Robertson, A., 158
Russell, L., 18
Russell, W., 18

Schauer, A., 64
Schmalhausen, I. I., 150
Schull, W. J., 18
Segregation distortion (SD) as a possible factor in irradiated populations of *Drosophila melanogaster*, 40
Segregational (genetic) load, 15
Selection
  density-dependent, 81–85
  frequency-dependent, 81–85
  hard, 81
  hard and soft, contrasting consequences of, 85–91
  soft, 80
  truncation, 101–103
Self-culling, 111–113
Shugart, H. H., 106
Simmons, M. J., 19, 50, 83
Simpson, G. G., 1
Singh, S. M., 118
Sober, E., 109
Soft selection, 80
  and the *corky* gene in cotton, 93–100
Somatic polymorphism, 122
Sonnenblick, B. P., 48
Spassky, B., 35
Stebbins, G. L., 1, 119
Stephens, S. G., 93
Stern, C., 41, 50
Substitution (genetic) load, 16
Sved, J. A., 100
Synthetic theory, 1

Temin, R. G., 41
Thoday, J. M., 107
Thompson, J. N., 65
Thompson, V., 25
Time arena, 151
Truncation selection, 101–103
Turelli, M., 119
Turner, J. R. G., 87, 135

U.N. Scientific Committee on the Effects of Atomic Radiation, 21
University, evolution of (an analogy), 2

Vann, E. G., 134
Van Valen, L., 114
Vetukhiv, M., 4, 64
Volterra-Lotka model, 83

Waddington, C. H., 4

173

Index

Wallace, B., 4, 24, 34–35, 40–41, 54, 62–66, 70, 80, 83, 93, 96, 105–106, 119, 134, 142, 145
Wallace-Kass model of gene control, 62
Westoby, M., 128
Williamson, M. H., 87, 135
Wilson, E. O., 4
Wood, K. V., 64

Woodruff, R. C., 65
Wright, S., 1, 8, 15, 66
Wright, T. R. F., 69
Wynne-Edwards, V. C., 145

Yoshikawa, I., 56

Zouros, E., 118